INCA COSMOLOGY AND THE HUMAN BODY

For Michel Despland,

with best regards,

Constance Classen

INCA COSMOLOGY AND THE HUMAN BODY

CONSTANCE CLASSEN

University of Utah Press
Salt Lake City

∞ This symbol indicates books printed on paper that meets the minimum requirements of American National Standard for Information Services—Permanence of Paper for Printed Library Materials, ANSI A39.38–1984.

LIBRARY OF CONGRESS CATALOGING-IN-PUBLICATION DATA

Classen, Constance, 1957–
Inca cosmology and the human body / Constance Classen.
p. cm.
Includes bibliographical references and index.
ISBN 0–87480–399–3
1. Incas—Religion and mythology. 2. Incas—Philosophy.
3. Incas—Rites and ceremonies. 4. Body, Human—Andes Region—Symbolic aspects. 5. Body, Human—Andes Region—Religious aspects.
I. Title.
F3429.3.R3C53 1993
299' .8—dc20 92–26079
CIP

CONTENTS

LIST OF ILLUSTRATIONS

PREFACE

My interest in the Incas grew out of my love for the folklore and music of northwestern Argentina. A decade ago I devoted two years to gathering folklore from family, friends, and books in the hills, homes, and libraries of the province of Tucumán, once part of the Inca empire. When I later came to study the history of the Incas, I was particularly pleased to discover that words integral to the Quechua-penetrated Spanish I knew had been integral to the language of the Incas, and that rites and customs I was familiar with in the mountains of Tucumán had been practiced centuries ago by the Incas in their capital, Cuzco.

While I was fascinated by the studies of Inca and Andean society I read, I was nevertheless often frustrated by the extent to which Andean culture was structured to conform to Western historical and ideological concerns. Thus, on entering the Doctoral Program in Religious Studies at McGill University in Montreal in 1986, I determined that for my thesis I would explore Inca religion within the framework of an indigenous model for understanding the world, that of the human body.

The present book is a revised version of that thesis. Certain portions of this work have appeared in Constance Classen, "Equilibrium and Exchange: The Relationship Between Land and Body in Andean Religion," *Journal of the Anthropological Society of Oxford* 18:3 (1987); "Sweet Colors, Fragrant Songs: Sensory Models of the Andes and the Amazon," *American Ethnologist* 17:4 (1990); and "Literacy as Anticulture: The Andean Experience of the Written Word," *History of Religions* 30:4 (1991).

I was very fortunate to have my research subsidized by doctoral fellowships from the Social Sciences and Humanities Research Council of Canada and the Fonds pour la Formation de Chercheurs et l'Aide à la Recherche of Quebec. I wish to express my gratitude to my thesis supervisor, Gregory Baum, for his unwavering support; to Henrique Urbano for his kind assistance during my time in Cuzco; to Domingo Bravo, under whose learned direction I studied Quechua (or Quichua, as it is known in Argentina) in Tucumán; to Anthony Synnott, Gilles Bibeau, Gary Urton, and Christine Franquemont for their comments and support; and to my editors at University of Utah Press for their assistance and enthusiasm during the process of publication. My special thanks go to David Howes, who shared with me his expertise on the anthropology of the body and the senses, and provided me with an inexhaustible supply of encouragement and advice.

Among my family members, I am grateful to my aunts and cousins in Argentina for helping me in my quest for Andean folklore, to my grandfather, for inspiring me with his stories of Tucumán in the early 1900s, to my brother George for his practical assistance with my research and writing, and above all to my mother, who has aided my work in so many ways.

Constance Classen
Center for the Study of World Religions
Harvard University, 1991

1
INTRODUCTION

THE INCAS

The Incas began as one people among many in the highlands of the Central Andes. By the early sixteenth century, however, after a period of conquest of about four centuries, they had built up an empire extending over forty-three hundred kilometers along the Andes from present-day southern Colombia to northern Chile and Argentina, encompassing a population of approximately 10 million.

The Inca empire, called Tahuantinsuyu (Four Quarters), was divided into four regions, each supervised by an imperial governor and a descending series of officials down to the level of the *ayllu*, a kin group holding land in common. At the head of the empire was the Inca (literally "lord"), who was held to be divine. Underneath him were the Incas in general (his blood relations), the Incas by privilege (important allies), the imperial administrators and local leaders, and the common people.

The Inca ruled from the capital city of Cuzco (in present-day Peru), the administrative and cultural center of Tahuantinsuyu. The sons of local leaders from throughout the empire were sent to Cuzco to be schooled in Quechua—the official language of the empire—and Inca traditions and skills, including the use of the quipu, a mnemonic device based on knotted strings. The Incas had no writing but kept extensive data, from traditions to statistics, stored on quipus.

The Inca state was based on agriculture, with herding of llamas and alpacas playing a secondary role. The principal crops were corn and potatoes. These were extensively cultivated through a well-planned system of terraces and irrigation canals. A portion of the harvest was kept each year in storehouses in case of necessity.

The Incas were able to control their vast territory in part by extending the deeply ingrained Andean practice of reciprocity of goods and services among individuals and communities throughout the empire. Each ayllu farmed a portion of its land for the state and the state religion, as well as contributing men for tours of duty in the army and state projects, and women for distribution as wives or *acllas*, cloistered temple attendants. In return, the state provided a measure of peace and order, public works, protection against famine, and a religious system.

The worship of the Sun, the mythical ancestor of the Incas, and of the Inca himself, was incumbent on the subject peoples of the empire. As long as they fulfilled their obligations with regard to the state cult, however, the Inca's subjects were free to retain their local beliefs. In turn, the Incas often incorporated elements of the cults of the peoples they conquered into their own religion.

In fact, Inca religion, like Inca government, was developed from long-standing Andean traditions. As in the latter, the major foci of worship were the ancestors and aspects of the natural world—the Sun, Thunder, the Earth—and the major rituals were based on the agricultural cycle and the human life cycle. All sacred places, beings, and objects were known as *huaca* (extraordinary and sacred). Each ayllu usually possessed one primary huaca, often a stone figure, that it worshipped as its founder.

The hierarchy of priests was ordered similarly to the administrative hierarchy, with a high priest (always a close male relative of the Inca) in Cuzco and a descending order of priests under him. The duties of these priests included caring for the temples, officiating at ceremonies, divining, and hearing confessions. Women could also hold certain sacerdotal positions: The cult of the Moon, the female counterpart of the Sun, for example, was presided over by women. As well, girls were chosen from throughout the empire to serve as cloistered acllas.

In 1525 the half brothers Huascar and Atahualpa, sons of the late Inca, began a bitter battle for control of the empire. Atahualpa won the contest in 1532 and assumed the position of Inca. It was at this point that the Spaniard Francisco Pizarro entered Tahuantinsuyu with a force of about 180 troops and, through a ruse, captured the Inca. Atahualpa was tried and executed by the Spanish after they had already collected an enormous ransom in gold for his release. The Spanish then proceeded to take control of the empire, and in 1572 Inca resistance ended with the execution of the last pretender to the Inca throne.

Many hypotheses have been put forth to explain how such a small force was able to take over such a large and well-organized empire, including the superior technology of the Spanish, the devastated state of the empire after the civil war, the presence of dissatisfied Inca subjects who were only too ready to aid the Spanish to overthrow their oppressors, and the initial Andean perception of the Europeans as divine. The most important factor, however, was probably the inability of the body of Tahuantinsuyu to act on its own once its head, the Inca, was captured. The social structure that had worked so well to integrate the empire into a hierarchical whole left the Inca's subjects unequipped to deal with the unpredictable audacity of the Spanish.

THE BODY AS SYMBOL

Our subject is how cosmology and the human body were interrelated by the Incas,[1] and how the Spanish Conquest was perceived by them according to the symbolic system derived from this interrelationship. In Inca cosmology the body served as a symbol and mediator of cosmic structures and processes through its own structures and processes. The fundamental structures of Inca cosmology—the dualities of right and left, high and low, male and female—were, in fact, derived from the structures of the human body. The processes of the cosmos, in turn, were modeled on the processes of the body—the intake and outflow of air and fluids, the digestion of food, the circulation of blood, reproduction, aging and death, and so on. Finally, as an integrated and dynamic whole, the body provided a model for the integrated and dynamic totality of the cosmos.

As the basic organizing metaphor of Inca cosmology, the human body therefore offers an indigenous paradigm for interpreting Inca religious practices and beliefs. The approach followed here is to first examine what the portrayal of the human body in Inca myths and rituals can tell us about the Inca model of the body, and second to see how the body model thus elaborated functioned to structure Inca symbolic life. By examining Inca cosmology through the indigenous model of the body, my aim is to produce a study of Inca religion that makes sense not only from a Western perspective, but also within the logic of Andean culture.

While the body model elaborated by the Incas was based on the physical model of the body, it was invested with meaning according to Inca ideology. One fundamental example of this is the Inca cultural construction of gender. In Inca cosmology the categories of male and female were not ultimately based on biological sex, but rather on cultural ideals. Males, for instance, were deemed to be dominant, and females subordinate. According to this scheme, a man in a position of subordination, such as a defeated warrior, was symbolically female. Thus to be a man or a woman in Inca culture it was not enough simply to be of the appropriate sex, one had to manifest the appropriate ideals.

If the basic principles of Inca cosmology were relatively simple, these could be developed in Inca myth and ritual in complex ways. Indeed, my primary task has been to discover how these basic propositions can be seen to underlie even the most apparently intricate and obscure Inca symbolic practices. By interpreting Inca religion through the Inca model of the body, many of the perplexing pieces of information we have concerning Inca symbolic practice—the Incas' burial of their foreign wives with earth from the wives' homelands, the association of ancestors with young trees, the conversion of the bodies of traitors into drums, the fasting from salt and pepper before rituals—begin to make sense.

At times the interpretation must remain inconclusive because of a lack of information. As the alternative, however, would be to offer no interpretation at all, I have chosen in such cases to at least present a tentative analysis. At the same time I hope that my explorations might stimulate further investigation in the modern Andes, where there is an abundance of data on the cultural role of the body potentially available.

In the last twenty years or so major works have been written on the sociology and anthropology of the body.[2] Two of the most important and influential are *Natural Symbols,* by Mary Douglas,[3] and *The Body and Society,* by Bryan Turner.[4] Here I do not deal specifically with the body as a symbol for society, as does Douglas, or with the socioeconomic regulation of bodies, as does Turner, but rather with the cosmological significance of the human body as expressed through Inca myth and ritual. For this reason, the extensive political and economic ordering of human bodies in the Inca empire—administrative divisions, systems of labor exchange, and so on—are dealt with only insofar as they exemplify cosmological principles.

In this book, therefore, I deal primarily with the role of the body in Inca religion and, while I draw on the insights provided by work done on the body and on religion in the social sciences, my research is undertaken within the field of religious studies. In this it is something of an anomaly. The study of the religions of nonliterate peoples has traditionally fallen to anthropologists, while students of religion have concentrated on those religions with texts.[5]

In recent years many anthropologists have moved into the study of the religions of literate peoples. There has not been a corresponding movement by scholars of religion into the study of the traditions of oral peoples. The scholar of Native American religions Sam Gill writes that he is

> confounded by why students of religion have so willingly neglected the study of Native American religions and, more widely, the religions of non-literate, or as I now prefer to say, "exclusively oral" peoples the world over. The material for the study of the religion of these peoples is rich, extensive, and available. . . . Yet despite the extensive records and current opportunities for study . . . the interpretative study of Native American religions has been virtually ignored by the modern academic study of religion.[6]

In keeping with Gill's direction, it is my hope that the work I undertake here will help to extend the study of religions to encompass oral cultures by presenting a model for understanding indigenous religions through their own organizing symbols, such as the body.

A topic not usually treated in cultural studies of the body, but that will receive special attention here, is sensory perception.[7] Our knowledge of the world comes to us primarily through our senses, and thus the value we assign to our different senses has a profound influence on the way we conceive of the world. In the modern West we understand the world primarily through visual paradigms, such as maps and charts, pictures and "perspectives," and these are the paradigms through which we tend to "see" other cultures as well.

The terms *cosmology* and *worldview,* for example, are often used interchangeably by anthropologists and other scholars. Is a cosmology, however, a "view"? A landscape of beliefs that can be accurately mapped out by someone with the correct vantage point? Or does the visualist model entailed in the use of this metaphor not gloss over the sensory and symbolic complexities inherent in any cosmology, as well as assume that all cultures assign the same priority to sight as the West?[8]

The way to correct for the tendency to perceive other cultures "through Western eyes"—that is, in terms of a Western sensory model—is to attempt to understand them through *their* sensory models. A start in this direction is evidenced in the work of many of the anthropologists of the 1950s and 1960s.[9] Claude Lévi-Strauss's analysis of the sensory codes in myths, for example, could have provided a basis for proceeding to the analysis of the sensory codes of whole cultures.[10]

In its growing preoccupation with hermeneutics, however, anthropology since the 1970s has instead tended to increase the visual bias in our perception of other cultures.[11] If religionists have concentrated on the study of sacred texts, many anthropologists have come to interpret culture itself as a text. This trend can be traced to Clifford Geertz's work, *The Interpretation of Cultures*, in which he defines culture as an "ensemble of texts . . . which the anthropologist strains to read over the shoulders of those to whom they properly belong."[12] The drawback to such an approach is not only that it reduces the multisensory dimensions of a culture to a flat, visual surface, but that it turns dynamic, interactive events into static, passive texts, which need only be properly "interpreted" by the ethnographer to be understood.

There are a number of anthropologists who have criticized the dominance of the visual in contemporary ethnography and have argued for the development of an "anthropology of the senses."[13] David Howes comments that "the ethnographer who allows his or her experience of some foreign culture to be mediated by the model of the text will have no difficulty in coming to think of the natives as enacting a particular 'interpretation' of the world in their ritual and other activities. But it would be more accurate to regard them as *sensing the world.* What is involved in 'sensing the world' is experiencing the cosmos through the mold of a particular sense ratio and at the same time making sense of that experience."[14]

To pursue this last point: in their myths and rituals the Incas manifest a distinctive way of sensing the world that is intrinsic to their cosmology. While sight, hearing, smell, taste, and touch are each assigned different characteristic functions in Inca cosmology, the Inca sensory model of the world is basically audiovisual. Sight is associated with the visible, structured past, situated in front of the body; and hearing with the dark, fluid future, situated behind the body. Thus in the central Inca temple of Coricancha there were both visual representations of the past—for example, depictions of the lives of past rulers—and oracles that spoke of the future.

For the Incas, a solely visual model could no more suffice to represent the cosmos than a solely male model would suffice to represent humanity. Sight needed to be complemented by hearing, just as masculinity needed to be complemented by femininity. For this reason, while the Incas gave symbolic priority to sight, it remained fundamentally allied to hearing and the other senses.

On one level, in fact, the combat between the Spanish and the Incas was a contest between two different sensory orders. This is particularly evident in the struggle for supremacy between the

Spanish and indigenous belief systems. An early seventeenth-century chronicle tells the story of an Andean caught between Christianity and his native religion. After the Andean was converted to Christianity, his ayllu's huaca tried to reassert its dominance over his senses by blinding him with light and making his ears ring. The Andean, however, proclaimed his faith in the new order, and the huaca lost the battle for control.[15]

Another Christian Andean of the same period makes it clear in a model prayer he composed for his fellow Andeans that Christianity is inseparable from a "Christian" sensory order:

> May Jesus Christ be in my eyes and in my sight. Cross.
> May Jesus Christ be in my ears and in my hearing. Cross.
> May Jesus Christ be in my nose and in my olfaction. Cross.
> May Jesus Christ be in my mouth, on my tongue, in all my words, in what I eat and what I drink. Cross.[16]

Despite centuries of Western influence on the Andean sensorium, however, the Andean way of sensing the world remains distinctive to this day. A striking example of the contrast between Andean and Western sensory orders (and of an act of sensory translation) was recently given in a talk titled "Sound Symbolic Discourse as Cinematic Text" presented by Janice Nuckolls at the 1991 conference of the American Anthropological Association. In that talk Nuckolls attempted to explain contemporary Andean auditory symbolism to a visualist North American audience by using cinematic metaphors such as "wide-angle shot," "close-up," and "slow motion."

THE SOURCES

The period covered here is from immediately before the Conquest until the end of Inca resistance in the late sixteenth century. The primary source of information is the large body of sixteenth- and seventeenth-century chronicles dealing with the Inca civilization. Naturally, these works are influenced by the cultural background and motives of their authors; nonetheless, they provide invaluable information on Inca traditions and institutions during a time when many of these were still extant, and memories of the pre-Conquest era still fresh.

The majority of these chronicles were written by Spaniards, many of whom were careful and dedicated Andean scholars. A few, however, were written or related by native Andeans. Of particular importance among these last is the extensive illustrated chronicle produced by the Andean Felipe Guaman Poma de Ayala, which was rediscovered only in 1908.[17]

There is a tendency among some contemporary Andean scholars, particularly those interested in "sifting" history from myth, to regard as false those Inca myths not compatible with assumed historical fact.[18] I am not concerned with such dichotomies; I am not attempting historical reconstruction but rather an examination of ideology. Accordingly, it is not of primary importance, for instance, whether Cuzco was indeed a model in miniature of the Inca empire, as the half-Inca, half-Spanish Garcilaso de la Vega states in his chronicle, or whether the literacy of the Spanish was actually a major factor in the Conquest, as Guaman Poma reiterates,[19] but that the Andeans *perceived* it as being so.

Andean chroniclers, such as Garcilaso and Guaman Poma, have been charged by Western scholars with presenting "ideal" rather than "factual" accounts of Inca culture. Inca culture, however, was itself ordered by ideals, by foundational myths and symbols. The very fact that the indigenous chroniclers did think and write in terms of ideals rather than facts, therefore, testifies to a basic Andean structure within their work. On this symbolic plane the indigenous chroniclers present us with a truer account of Inca culture than a straightforward factual description could do.

The Andean chroniclers, along with the Spanish, were, of course, under pressure to situate the Inca material within a Western Christian framework, and this produced certain distortions in their accounts of Inca culture. At times the introduction of Western elements in an Andean narrative is obvious, at other times less so. The one check for such distortions is to compare a particular account to other related available data and determine whether it makes sense within an Andean framework. Fortunately, the abundance of information on many aspects of pre- and post-Conquest Andean life in the chronicles usually makes such checks possible, and, indeed, most of the data on the Incas used here is seconded by other period sources.

There are a number of excellent studies on contemporary Andean cosmology and ritual, and I have used these to supplement the material found in the Conquest chronicles and relate it to the present day. While it would be naive to assume that modern Andean culture has survived unchanged from the time of the Incas, it would be equally misguided to disallow any cultural continuity from that period to this, considering the traditionalist nature of Andean society and the many similarities between modern Andean culture and Inca culture. Thus, while modern Andean material cannot be used as "proof" of the existence of an Inca belief or practice, it can suggest ways in which the latter may be interpreted and understood, as well as demonstrate the fundamental significance of certain principles to Andean culture.

Of greatest relevance to our topic here is the research done by Joseph Bastien on how the human body functions as an indigenous model for ordering society, land, and cosmos in the modern Andes. The Andean people Bastien focused on are the Bolivian Qollahuayas, who exercise their influence throughout the Central Andes as traveling medicine men. Bastien's research makes plain the importance of body symbolism in modern Andean cosmology.[20]

Little work has been done on sensory symbolism in Inca or Andean culture, although some recent studies in the field show an interest in the topic.[21] Of particular value are Veronica Cereceda's thought-provoking exploration of Andean aesthetics and Regina Harrison's analysis of Andean oral and visual codes.[22] With regard to the Inca and colonial periods, R. T. Zuidema has examined the role of sensory perception in certain Inca rituals.[23] Finally, Billie Jean Isbell and Lawrence Sullivan have undertaken comparative studies of the metaphorical role played by the senses in various South American cultures, including that of the Andes.[24]

I mean *Inca Cosmology and the Human Body* to complement and expand on the work done by these authors by examining what the corpus of Andean material tells us about the interplay of cosmology and the human body among the Incas, and exploring how the model of the body thus revealed ordered Inca culture. I do not by any means pretend to have written an exhaustive or conclusive study of the subject. My aim is, rather, to present an analysis of Inca body symbolism based on the available data and to suggest topics for further research—specifically among the peoples of the Andes, who undoubtedly have much more to offer in this regard than has hitherto been attended to, but also in any culture accessible to a similar analysis.

The use of the human body as a cosmological symbol is a worldwide phenomenon, as has been documented in a multitude of studies,[25] and a basic part of Western heritage.[26] An examination of the Incas' comprehensive and involved use of such symbolism can provide a model for understanding and comparing similar symbolic systems in other societies, as well as a counterpart to the symbolic ordering of the body in modern urban culture.

In *The Conquest of America: The Question of the Other* Tzetan Todorov writes that "The discovery of America, or of the Americans, is certainly the most astonishing encounter of our history. We do not have the same sense of radical difference in the 'discovery' of other continents and of other peoples."[27] In this book I offer both an alternative to the Western worldview in the cosmology of the American "other," and a look at how that worldview was perceived by the people it conquered and marginalized.

2
COSMOLOGY

EQUILIBRIUM AND EXCHANGE

For the inhabitants of the austere Andean region to survive and prosper, they had both to maintain a careful equilibrium between themselves and their environment and to establish a system of exchange among communities living at different altitudes, with access to different resources.[1] This concern for equilibrium and exchange was also manifested in Andean cosmology, as elaborated in the myths and rituals of the Incas and in contemporary Andean religion. Isbell says of the Andeans of Chuschi, Peru, that "one of their principal concerns is to maintain equilibrium and order in a world in which dual forces are balanced against one another."[2] Enrique Oblitas Poblete writes that the Callawayas (or Qollahuayas) of the Bolivian Andes conceive of the universe as held in balance by the opposing forces of existence and nonexistence (*mana cajmanta caj*), one of which strives for the stability of matter and the other for its disintegration.[3]

These forces, which can alternately be seen as structure and fluidity, clarity and obscurity, fertilizing and fecund, should not be equated with good and evil, although they can be understood as such in certain contexts, for each is necessary to complement and counterbalance the other. A Bolivian miner expresses this complementarity in one instance: "We think of the moon as the force that generates cold. If only the moon existed, we would all die, frozen by the cold. Well, if only the sun existed, and not the moon, who knows but what we would all die, burned to a crisp. And so there is this belief that these two worlds give us an equal temperature."[4]

A mere equilibrium of forces, however, results in a static universe. In order to integrate and animate the cosmos, a system of exchange is needed. This combination of equilibrium and exchange is expressed by the Quechua word *ayni*, meaning balance and reciprocity.[5] In the words of one Peruvian Andean, "the whole universe is *ayni*."[6] A primary function of Andean religion is to set out the basic divisions of the cosmos, maintain them in harmony, and establish a means of exchange among them. In this endeavor the human body, through its structure and functions, serves both as a model of and for the cosmos, and as a mediator among cosmic forces.

BODY STRUCTURES

The most basic expressions of Andean dualism—male/female (*urco/china*), right/left (*paña/lloque*), high/low (*hanan/hurin*), external/internal (*hahua/ucupi*), and so on—originate in the structure of the human body (*ucu*).[7] The concern of Andeans for these bodily oppositions is witnessed in many of their terms for body parts. Domingo de Santo Tomás's Spanish-Quechua dictionary of 1560, for example, gives separate words for the upper lip, *virpa*, and the lower lip, *cirpi*; and for the upper ear, *cotcolli*, and the lower ear, *lluto*.[8] (For a compilation of colonial and modern Quechua body terms, see the Appendix. Because there is no standard Quechua orthography, the spelling of Quechua words will vary in accordance with the source used.)

These antonymous expressions tend to be linked with each other as follows: male, right, high, external; and female, left, low, internal. The first set is generally associated with structure, clarity, and fertilizing power; and the second with fluidity, obscurity, and fecundity, as will be seen in the analyses of Inca and Andean symbolism that follow. These characteristics are extended to apply to virtually everything in the cosmos. The sun, *inti*, for example, identified with structure, clarity, and fertilizing energy, is a classic male symbol in Andean thought, while the moon, *quilla*, linked with fluidity, obscurity, and fecundity, is a classic female symbol. In Andean drawings the sun is typically placed to the right, and the moon to the left (from the perspective of the drawing, not the viewer).

These different contrasts are combined in the following passage, written by a Spaniard in 1585, describing the division of Collasuyu, the southern quarter of Tahuantinsuyu, into two moieties, Urcusuyu and Umasuyu. "[The Urcusuyus] dwell in the mountain highlands, called *urcu*, and the Umasuyus on the flat lowlands beside rivers and lakes, called *uma* in this language [Aymara]. Some say that Urcusuyu means a vigorous and manly people because urcu denotes masculinity; and umasuyu feminine and of less worth. The Urcusuyus were always prouder and of better quality, and the Inca placed them at his right hand in public places and they were more highly regarded than the Umasuyus."[9] In this one quotation we find the qualities of high, dry (structure), right, male, and superior associated; and those of low, wet (fluidity), left, female, and inferior associated.

Although male is usually understood as being superior to female, right superior to left, and so on, in Andean thought, each is essential to the existence of the other. There can be no body without both right and left sides, and no normal adult body is complete without a partner of the opposite sex. The term in Quechua for pairs, such as male and female, or eyes or hands, is *yanantin* (helping together), while *chulla* is the word for the single half of what should be paired; for example, *chulla ñaui* (one-eyed). Tristan Platt writes of the Macha Indians of Potosí, Bolivia, that "it appears, then, that the human couple is conceived in the same terms as the symmetry of the body. . . . Men and women ought to be *yanantin*—that is, they should partake of that perfect union achieved by the two halves of the human body."[10] This concept of the conjugal pair as a fundamental unit is expressed by the term *karihuarmi* (man-woman) The human body, thus, in itself or paired in complementary opposites, is a basic symbol of wholeness and integrity in the Andes.

To avoid confusion or the appearance of contradictions, it must be stressed that all of these antonymous characteristics are relative to their context. Water, for example, which is generally conceptualized as female, can become male when it is seen as semen fertilizing the earth. The Earth, in turn, also generally conceptualized as female, can become male when it represents structure as opposed to the fluidity of water. Even human bodies are not unequivocably one sex or another. The Incas, for instance, would send women's clothes to men they deemed to have failed in their masculine role.[11]

Male and female here are cultural constructs, and not solely based on physical differences. Bodies acquire their identity in relation to other bodies. Furthermore, while a body, organic or otherwise, as a whole may be considered male or female, it can contain elements associated with the opposite sex. For instance, just as the male human body can have aspects that are associated with femaleness (e.g., the left side), so can the male sky have female elements (e.g., the moon).

The final type of bodily oppositions dealt with here are inversions: front/back and rightside up/upside down. In Andean thought the past is situated in front (*ñaupa*) and the future behind (*quepa*). English parallels are the words *before*, which means both "in front" and "earlier," and *after*, which means "spatially behind" and "temporally ahead." Diego de González Holguín, in his dictionary of 1608, gives the verbs *ñauparina* (to go in front) and *ñaupacchani* (to be early); and *qqueparini* (to go behind) and *qquepachanan* (to be late). The noun *ñaupaqquey* means both "predecessor" and "one who is in front." The future, thus, is not something that one can walk ahead into, but rather, something that one has to turn oneself around(*tikraricuni*) to reach. The Incas, in fact, called the cataclysmic event they considered to divide one age from the next *pachacuti* or *pachaticra*, reversal of time and space.[12]

The words for "upright" and "standing" are *ticnuni* and *sayani.* The former is related to the word *ticci* (zenith), and the latter to *sayay* (height and personal presence). The term *sayak sonco* (upright heart) means "to be firm and constant." Uprightness is therefore symbolic of order and morality. The word for "upside down" is *tiksu,* which can also mean "disarranged" and "nadir." To be hung upside down was a common punishment for crime in the Inca empire.[13] Invertedness, therefore, symbolizes disorder and immorality.

It is not only the oppositions of the body that are important in Andean culture, but also the "in-between" parts of the body. On the most obvious level, for example, the nose, *cenca,* can be understood as mediating between the eyes, *ñaui,* and the mouth, *simi.* In Quechua, however, we also find terms for the space between the nose and the mouth (*simi pata*), the area between the shoulder blades (*wasa waycu*), the furrow between the chest and the stomach (*q"asqu puxyu*), and so forth.[14] These areas serve both to divide and to mediate, and the fact that they have separate names demonstrates the importance of in-between spaces and dividing lines for the Andeans.[15]

The word for split in relation to body parts, huaca (based on huac: double, other, or apart), is, in fact, the same as the word for a holy place or object. This may have to do with the concept of a sacred place as a crack or mediating space between the supernatural and the natural, or else as something that exists on both levels at once and thus is double. In any case, the use of huaca to mean "dangerous," such as in the term *huaccap ñan* (dangerous pass or place), indicates that this liminal zone between categories was a place of hazard for Andeans.

The structure of the human body, of course, is not static but changes throughout life and can be altered by external circumstances, such as disease or mutilation. Nor are all human bodies identical in form. The Andean concern with physical differences is witnessed by the extensive number of Quechua terms characterizing different body parts; for example, *cumu cenca* (curved nose), *quimray cenca* (wide nose), *chusna cenca* (small nose), and *vecro cenca* (twisted nose).

The Andeans had an ambivalent attitude to physical abnormalities. On the one hand, they were considered the manifestation of a moral defect, or at least unpropitious. During the most important Inca rituals, for example, all persons with physical defects were sent out of the capital.[16] On the other hand, abnormalities were sometimes considered a manifestation of the divine. This was particularly true of duplicated or divided body parts or bodies; for example, *huaca virpa* (split lip), *huaca runa* (person with six fingers), or *huaca huachasca* (twin). The sixteenth-century Andean Joan de Santacruz Pachacuti Yamqui, gives the definition of huaca, in relation to the sacred shrine of Pachacamac on the Peruvian coast, as "a split, very ugly, or flat nose"![17]

In such cases it is significant that the corporeal division was often thought to have been effected by lightning, a potent transforming agent and carrier of divinity.[18] The reason for the ambivalence the Incas manifested toward these physical abnormalities is probably that they were seen as both a deviation from the sanctified order and a revelation of the transforming power of the divine.

BODY FUNCTIONS

The structure of the body is inseparably linked to its functions. Bastien writes that the Qollahuayas of Bolivia understand the body as a hydraulic system, circulating and distilling fluids and semifluids (water, air, blood, and food). Illness is a disruption of this cycle, causing a gradual "drying up" (weakening, wasting away) of the body. Death is the permanent separation of the wet and dry elements of the body.[19] "Wholeness (health) of the body is a process in which centripetal and centrifugal forces pull together and disperse fluids that provide emotions, thoughts, nutrients, and lubricants for the members of the body. Moreover, this process extends beyond the dualistic confines of inner and outer, in that fluids of the body are governed by similar dynamics within the environment."[20] Body fluids, like rivers, must be properly controlled so that they integrate the structure of the body (the land) without disintegrating it.

We do not know whether the Incas held similar notions of health and illness; however, it is interesting to note that the Qollahuayas served as medical and religious advisors to the Incas. In González Holguín we find the expression *chaqueymamani*, "to dry up slowly with illness," which suggests a similar understanding of the role of body fluids in the Andes of the sixteenth century. Bastien sees a parallel to the Qollahuaya concept of the body as a hydraulic system in the emphasis on the environmental circulation of water in Inca myths.[21] Certainly the hydrologic cycle was the subject of extensive symbolic elaboration by the Incas.

In Inca and modern Andean thought the middle section of the body contains the organs that control the flow of fluids through the body. It serves, therefore, not only as a mediator between top and bottom, but as a center of exchange. These central organs are indicated by the Quechua term *sonco*, which refers to the "entrails": heart, stomach, and intestines. The sonco is held to be the seat of emotion and, together with the head, of thought.[22] As the center, the sonco both integrates the body into a whole and stands for the whole. The term *sonco*, in fact, could be used to mean "person," as in the expression *huac sonco* (double heart), "a duplicitous person."[23]

With the presence of the integrating and separating center, Andean dualism becomes tripartite. The number three, in fact, repeatedly comes up in Andean myth and ritual: Major Inca divinities such as the Sun and Thunder had three manifestations; Inca kin groups were related to each other in threes; and the cosmos itself was divided into three parts—the Upper World, This World, and the Lower World.

The second category of body functions covered here (leaving aside other kinds of body functions to be examined later) is that of sensory perception. The importance of acute sensory perception for the Andeans is demonstrated by the following entries in González Holguín:

> *Ccazccaruna,* one who uses all his senses sharply
> *Ccazccaylla ccazcachini ccazccachini ricuyta,* or *ccazccactam ricuni,* to see subtly
> *Uyariytam ccazccachini,* or *ccazccactam uyarini,* to hear subtly
> *Mutquiyta,* to smell [subtly]
> *Malliyta,* to taste [subtly]
> *Llancayta,* to touch [subtly]
> *Yuyayta unanchayta,* to imagine or understand subtly.[24]

Similarly, terms are listed for the loss of the different senses; for example, *ccaymarayak*, "one whose sense of taste is damaged, who cannot taste food."[25]

González Holguín gives no general Quechua term for the senses, and there possibly was none. In his seventeenth-century dictionary of the closely related Aymara language, Ludovico Bertonio specifically states that there is no generic term for the senses in Aymara.[26] There is a term for the senses in modern Quechua, however: *cawllaycuna* (see the appendix).

The senses for which González Holguín provides Quechua terms are sight, hearing, smell, taste, touch, and intelligence. In this last case González Holguín may have been influenced by the contemporary Western model of the senses, which counted intelligence as an "interior sense."[27] However, it may have been that the Incas themselves considered thinking analogous to seeing or smelling.

No Quechua terms are provided for proprioception or kinesthesia; however, these sensory phenomena were certainly not without symbolic significance for the Incas. In one early chronicle we read, for example, that the Andeans believed that twitches, buzzing of the ears, and similar sensations were omens of one thing or another. "When their eyelids or their lips tremble, or their ears ring, or any part of their body trembles, or they stumble, they say they will see or hear something good or bad: good if it is the right eye, ear, or foot; bad if it is the left."[28]

The senses given the most attention in myths and rituals were sight and hearing. Inca prayers, for instance, often express a desire to see and hear the Creator and to be seen and heard, as in the following excerpt:

Ah Viracochanticcicapac. . . .	O Viracocha, lord and founder. . . .
maypincanque	Where are you?
mana choricayquiman	May I not see you?
hananpichum	Above?
hurimpichum	Below?
quinraynimpichum	Beside?
capacosnoyqui	Where is your throne?
haynillabay. . . .	Oh, speak to me! . . .
camman	I turn to you
allcananiyvan	with my faulty eyes.
riacytam munayqui	I wish to see you,
ricuptiy	because by seeing you,
yachaptiy	knowing you,
unanchaptiy	recognizing you,
hamuttaptiy	comprehending you,
ricucanquim	you will see me,

yachavanquim	you will know me.
inticaquillaca	The sun, the moon,
ppunchao catutaca	day, night,
pocoyca chiraoca	the time of harvest, the time of cold,
manamyancacho	not in vain
camaschiscam	are they ordered,
purim	they go
unanchascaman. . . .	according to signs. . . .
haynillavay	Oh, speak to me,
oyarillabay	hear me,
manaracpas saycoptiy	don't let me grow tired
uañuptiy	and die.[29]

Harrison sees this prayer as displaying an Andean order of perception, with sight (*ricuna*), while the highest of the senses, providing the most superficial form of knowledge, and comprehension (*hamuttay,* which has connotations of reasoning and ordering) the most profound.[30] The Quechua verb translated here as "recognizing," *unanchani,* also has visual associations, as it means both "to consider and to understand" and "to make signs and to trace."

Sight in this prayer seems to be related to the apprehension of categories: above/below; sun/moon; day/night and so on. Speech and hearing, in contrast, are related to the life force: "Speak to me, hear me, don't let me grow tired and die." As well, there is the notion of a reciprocity of perceptual impressions ("by seeing you, . . . you will see me"), which is a further manifestation of the importance for the Andeans of exchange on all levels.

Earlier we saw that the front of the body is associated with the past and the back with the future in Andean thought. The reason for this is that the past is known and therefore visible (in front of one), and the future is unknown and therefore invisible (behind one).[31] The Quechua word for eye, ñaui, is, in fact, a synonym for "front," while the Aymara word for eye, *nayra,* also means both front and past.

If sight is connected with the front of the body, hearing seems to be associated with the back: One can see only in front of one, but one can hear sounds from behind one. In fact, in the Andes sound and darkness are often ritually associated.[32] In support of this theory, González Holguín gives as the definition of *qq̧uepa* both "trumpet" and "behind."

Sound can serve as a vehicle of transition in the Andes. Gary Urton writes of a contemporary Peruvian community that "the greatest amount of noise in the village occurs at dawn and at dusk:

unnecessary noise, even that of animals, is not easily tolerated at any other time. . . . Noise marks the transition from day to night and from night to day, it is inappropriate at other times."[33] Here noise represents the disorder of the liminal period in between night and day and also serves to mediate between one stage and another. Sound played a similar role in Inca rituals of transformation.[34]

In its integrating and transforming capacity, sound can be compared to a fluid. As with fluids in Andean thought, sound is exchanged between the body and the environment. For example, in the Inca ritual of *citua*, people would symbolically throw their misfortunes into the river while shouting, "Illnesses begone!" This symbolic fluidity of sound therefore associates it with the fluidity of the future, while sight is connected with the past because of the latter's relatively structured state. The past and present are what one knows and sees, the future can be heard of only through oracles.

The association of sight with structure, and hearing with fluidity, in Inca cosmology suggests a further association of the former with masculinity and the latter with femininity. The word *eye* (ñaui in Quechua, nayra in Aymara) was used in the Andes to mean "a sharp point," or a "sprout" or "seed," which terms have male connotations. Isbell reports that the inhabitants of Chuschi, Peru, in fact use the expression "father eye" to refer to the springs that impregnate the earth.[35] The ear, in contrast, as a dark opening, has female connotations of obscurity and fecundity. The ear-piercing ceremony of Inca youths, for example, served in part as a fertility rite.[36] Sight is considered an act of force by the Andeans. The expression *huañuytam ccahuaccuni* means to "kill with a look." Hearing is an act of submission: The verb *to hear* is *uyarini,* and the verb *to obey, uyani.* A *huanco rincri* (hollow ears), on the other hand, is "a disobedient person who does not hear what is ordered."

If sight, as male, and hearing, as female, have contrary associations in Inca thought, however, in Inca myth and ritual they are often intimately linked, as we shall see in the following chapters. This sensory integration seems to be the perceptual equivalent of *karihuarmi* (man-woman): sight-hearing, expressing a basic complementary unity of the contrasting senses.[37]

PACHACUTI YAMQUI'S DIAGRAM OF INCA COSMOLOGY

In his "Relación" of 1613, the native Andean Pachacuti Yamqui[38] drew what he said represented the cosmological model inside Coricancha, the central Inca temple in Cuzco. Whether this model actually existed in Coricancha is not known. What is significant, however, is that Pachacuti Yamqui held his diagram (figs.1 and 2) to be so central to Inca cosmology as to situate it in the central temple of the Incas.

The man and woman standing in the middle constitute the focal point of the diagram, illustrating the centrality of humans to the cosmos and the essentiality of both sexes. A single human figure would not have adequately conveyed the dual nature of the cosmos.

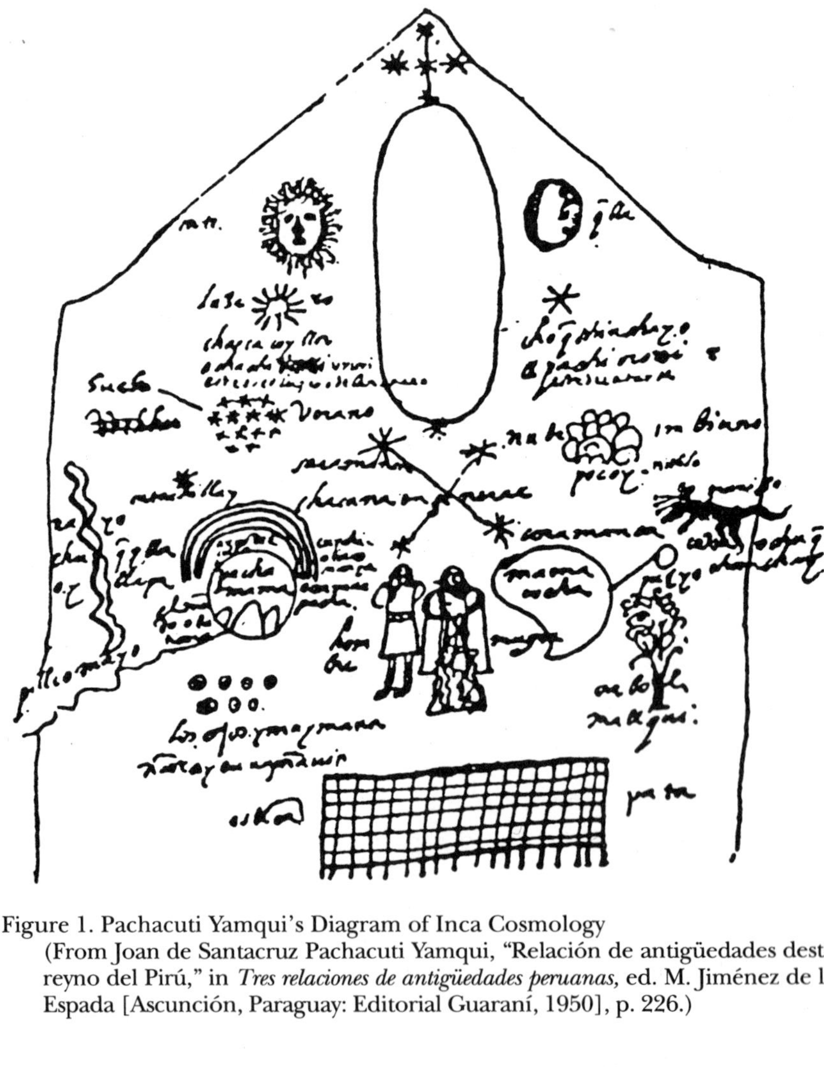

Figure 1. Pachacuti Yamqui's Diagram of Inca Cosmology
(From Joan de Santacruz Pachacuti Yamqui, "Relación de antigüedades deste reyno del Pirú," in *Tres relaciones de antigüedades peruanas,* ed. M. Jiménez de la Espada [Asunción, Paraguay: Editorial Guaraní, 1950], p. 226.)

Aside from the two humans and the *choquechinchay* cat (a mythological feline, included here because of its supposed role in producing hail)[39] to the right, there are no other bodies depicted in the diagram. This draws attention to the unique place of the human body in the cosmic order.

The top part of the drawing contains celestial phenomena; at the bottom is a representation of a storehouse and agricultural terrace (*collca pata*), representative of the Earth. The upright human figures in the middle act as mediators between the earth and the

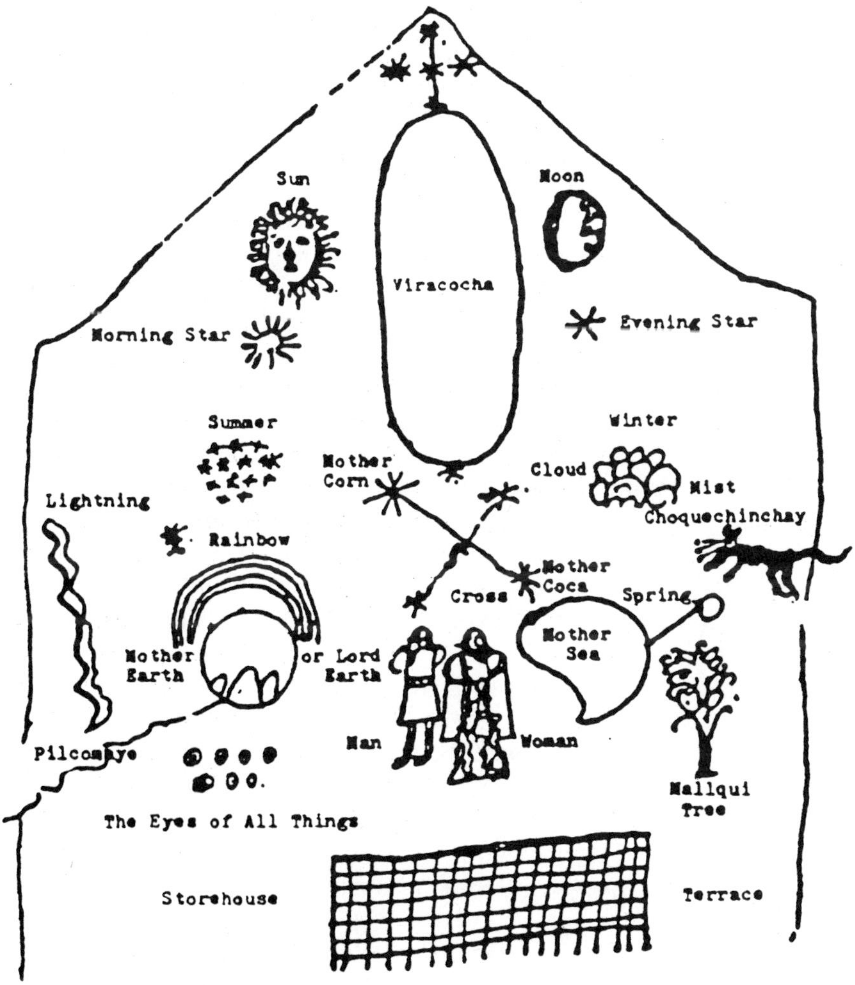

Figure 2. Pachacuti Yamqui's Diagram, with Labels Translated

sky, an axis through which power can flow. The drawing's vertical center, which can perhaps be compared to the dividing line between the right and left halves of the body, is, in fact, taken up by symbols of conjunction: the human couple; the stellar crosses; the grid of the terrace; and the oval, which Pachacuti Yamqui identifies as Viracocha, Creator of the World.[40]

In reference to the symbolic representation of Viracocha in the diagram, Pachacuti Yamqui writes, "whether it be male or female."[41] If the human couple in the center represents the basic concept of karihuarmi in this world, Viracocha can be said to embody the ideal original integration of male and female. This union of male and female was also expressed in hermaphroditic sculptures worshipped at various sites in Tahuantinsuyu.[42] The oval signifies Viracocha's transcendence of dichotomies, and inscrutability. It represents unbroken oneness in contrast to the diversified and structured cosmos represented by the grid of the terrace at the opposite end of the diagram.

It is out of this original unity that the division of the cosmos proceeds: male elements, on the left (pictorial right); and female elements on the right (pictorial left). The sun heads the male elements and the moon the female ones. In this respect Pachacuti Yamqui's diagram is an example of Andean parallel descent—men descending from men and women from women, with the sun and moon at the top as the first parents.[43] Within an Inca imperial context this gender parallelism also applied to social institutions: The Inca was at the head of male institutions under the patronage of the Sun; and the Inca's consort, the Coya, was at the head of female institutions under the patronage of the Moon.[44]

The significance of this male/female dualism in Inca cosmology merits emphasis. According to Andean gender ideology, male and female were complementary and independent categories, each with its own parallel sphere of concern. A practical instance of this gender parallelism is the Andean custom that women inherit property from their mothers, and men from their fathers. This custom, and the deep-seated structures that underlie it, made it impossible for post-Conquest Andean women to understand why, according to Spanish law, their property should now be controlled by their husbands.[45]

In Pachacuti's diagram the elements on the male side, on the whole, manifest light and dryness: the sun; the morning star ("grandfather"); lightning; the rainbow (which appears when the rain has stopped); and the Earth. Pachacuti labels this last "pachamama o camac pacha" (Mother Earth or Lord Earth), as though indicating that, although the Earth is usually female, here, in relation to the opposing Mother Sea, it is male.[46] The Pilcomayo River, shown flowing out of the Earth, is probably included here because of certain mythological associations it had for the Incas. Those elements on the female side of the diagram are associated with obscurity and wetness: the moon; the evening star ("grandmother"); the cloud; fog; hail; the spring of water; and the sea.[47]

This division between right and left, male and female, cannot be glossed as one between sacred and profane,[48] for all the elements on both sides of the drawing had religious significance for the Incas. The distinction is rather between kinds of sacredness. The items on the male side—lightning, transmitter of divine power; the rainbow, associated with the mediating role of the Inca; corn, the most sacred of foods—have a closer association with spirituality and the Upper World, while those on the female side—the supernatural choquechinchay cat; the *mallqui*, ancestor tree; coca, used for divination—are more closely related to divination and the under world.

The middle section contains all the symbols of culture in the diagram: the storehouse and terrace, ground of civilization; the clothed human couple, basis of society; domestic plants and animals symbolized by stars (the uppermost constellation represents llamas[49]); and Viracocha, the original giver of culture. Culture for the Incas was therefore an integration of male and female forces.

This understanding of culture exists in the modern Andes as well. Olivia Harris writes of the Laymi of Bolivia that "it is the fruitful cooperation of woman and man as a unity that produces culture, and this is counterposed to an unmarried person as non-cultural; culture is based on duality and contrasted with what has remained single when it should be paired."[50]

The right-left pairing of opposites—sun/moon, Earth/sea— in the drawing suggests that the Eyes of All Things and the mallqui tree, as the last items on either side, are meant to form a pair of complementary opposites.[51] It is difficult to determine exactly what the Eyes of All Things are meant to represent. According to Andean symbolism, they could stand for springs of water, seeds, or stars.[52] Springs are regarded as places of origin for animals and humans in Andean thought, seeds are the embryos of life, while stars are the prototypes of all terrestrial beings.[53] In all of these cases one has the image of the Eyes of All Things as a repository of ideal forms.

González Holguín defines mallqui as a "young plant" or "fruit tree." Mallqui could also mean "the body of an ancestor."[54] In Andean thought, therefore, a tree could apparently stand for the human body. Pachacuti Yamqui gives two instances of this. He writes that the first Inca revered his parents in the form of two trees from which he ordered golden fruit be hung, saying that the future Incas would be as the fruit of these trees (see fig. 5). Further on he describes the case of a wounded Inca warrior who laments that he is going to die without having produced any fruit. The warrior's body is then buried in the trunk of a tree, which produces fruit with healing powers.[55]

In Pachacuti Yamqui's diagram, consequently, the living humans in the center can perhaps be understood as divided into ideal forms, the Eyes of All Things, on the male side; and physical form, the mallqui, on the female. The Eyes of All Things, as the original ideal forms, could then also represent the past, and the mallqui, as the fruit-bearing tree or procreator of forms, the future.

In any case, the fact that the Eyes are apparently drawn as *eyes*, rather than stars or springs, along with the emphasis on light on the left side, certainly implies an association of sight with masculinity. There is nothing specifically indicative of hearing on the right side (unless one wishes to use one's imagination and take the Earth for an eye and the sea for an ear); however, the darkness and fluidity of the female half, particularly when contrasted with the opposing symbols of light and sight, could allude to aurality. In this regard it is important to note that if Pachacuti's diagram was indeed in the Inca temple of Coricancha, its visuality would have been complemented by the orality of the huacas housed in the temple.

Finally, while the diagram is a model of sacred reality, it is also that of an agricultural system and the relation of human beings to nature. The alternating seasons, the cultivated earth, the stars that indicate the times for sowing and harvesting are all essential features of agriculture. Of foremost importance in this system is the flow of water. Both sides of the diagram contain dynamic elements symbolic of meteorologic forces—the lightning that strikes the earth from the sky; the rainbow that originates in the ground, reaches the sky, and touches the ground again; the choquechinchay cat that rises up with the mist and casts down hail—that connect the sky with the earth and make agriculture possible. It is this fundamental circulation of water, up from the earth and down from the sky, that unites the cosmological model in a total dynamic.[56] Once again, therefore, as in the Qollahuaya concept of the body as a hydraulic system, we have the image of a structure animated and integrated into a whole by an exchange of fluids.

3
CREATION

THE VIRACOCHA COSMOGONY

The natives of this land say that in the beginning, before the Earth was created, there was someone called Viracocha. He created a dark world without sun, moon, or stars, and because of this creation he was called Viracocha Pachayachachi, which means Creator of All Things. After he created the world he formed a race of disproportionately large, painted or sculpted giants to see whether it would be good to make men that size. As they seemed much larger than himself, he said, "It is not good for people to be so large; they had better be my own size." Thus he created men in his likeness, as they are now. And they lived in the darkness.

Viracocha ordered these people to live in peace and to know and serve him. He gave them a precept to follow, with the warning that if they did not, he would confound them. They kept this precept, which is not specified, for some time. However, the vices of pride and cupidity were born among them and they broke the precept of Viracocha Pachayachachi, incurring his wrath, and he confounded and cursed them.

Then some were turned into rocks and others into other forms. Some were swallowed by the Earth and others by the sea, and over everything he sent a flood, which is called *uno pachacuti*, which means "water that overturned the Earth." They say that it rained sixty days and sixty nights, and that all of creation was flooded. There remained only a few signs of those who had turned into stone to serve as a remembrance of the event and an example to posterity. . . .

When Viracocha Pachayachachi destroyed the Earth, he kept three men with him, one of whom was called Taguapaca, in order to serve him and help him create the new peoples who would be made in the second age, after the flood. This was done in the following way.

After the flood was over and the Earth dry, Viracocha decided to repopulate the Earth, and in order to do it more perfectly, he decided to create luminaries that would give light. To do so he went with his servants to a great lake, which is in Collao, and in this lake there is an island called Titicaca. . . . Viracocha went to this island and ordered the sun, moon, and stars to emerge and go to the sky to give light to the world, and so it was done. It is said that he created the Moon with more light than the Sun, and that the Sun, out of envy, threw a handful of ashes in the Moon's face as they were about to rise to the sky so that it darkened to the color it has now. . . .

Taguapaca disobeyed the commandments of Viracocha when the

latter was giving some orders to his servants. Viracocha, angry at Taguapaca, ordered his other two servants to bind him head and foot and put him in a boat in the lake, and so it was done. As Taguapaca blasphemed against Viracocha for what he had done to him, and threatened to return and take revenge on him, he was carried away by the water that drained the lake, and he was not seen again for a long time. This done, Viracocha raised a solemn huaca in that place as a sign of what he had created and done there.

Leaving the island behind, [Viracocha] passed over the lake to the land, taking with him the two servants he had retained. He went to a place, now called Tiahuanaco in the province of Collasuyo, and in this place he sculpted and drew on some large stones all the nations he thought to create. This done, he ordered his two servants to commit to memory the names he told them of the peoples he had painted and of the valleys and provinces and places from where they would emerge, which were those of all the Earth. He ordered each of them to take a different route and call the above-mentioned peoples and order them to come out, procreate, and swell the Earth.

The servants, obeying Viracocha's command, set themselves to the task. One went along the sierra of the coast by the southern sea, and the other along the sierra of the awesome mountains we call the Andes, to the east of that sea. Along these sierras they walked, calling out: "O peoples and nations! Hear and obey the commandment of Ticci Viracocha Pachayachachic, who orders you to come out, multiply, and swell the Earth!" And Viracocha himself did the same along the lands in between those of his two servants, naming all the nations and provinces he passed through. On hearing the calls, every place obeyed, and some people came out of lakes, others out of springs, valleys, caves, trees, caverns, rocks, and mountains, and swelled the Earth and multiplied into the nations that are in Peru today. . . .

Another strange tale is also told. After Viracocha created all the peoples he arrived at a place where a group of men created by him had gathered. This place is now called the town of Cacha. When Viracocha arrived there the inhabitants found his dress and conduct strange. They murmured against him and plotted to kill him from a hill. As they took up arms for this purpose their bad intentions were perceived by Viracocha. Kneeling on the plain and lifting his hands and face up to the sky, he caused fire to fall from above on the hill. The whole place was set on fire, the land and rocks burned like straw. As those bad men were afraid of that terrible fire, they came down from the mountain and threw themselves at Viracocha's feet, asking to be pardoned for their sin. Moved with compassion, Viracocha went and put out the fire with his staff. . . .

Viracocha continued along his route, doing his works and instructing the people he had created. In this way he arrived at the district that is now Puerto Viejo and Manta on the equinoctal line, where he joined up with his servants. Desiring to leave the land of Peru, he gave a talk to those he had created, telling them of things to come. He said that there would come some people who would say they were Viracocha, their Creator, and that they should not believe them, that in the times to come he would send them his messengers to protect and teach them. This said, he and his two servants went into the sea and walked over the water as though on land, without sinking. Because he walked over the water like foam, he was called Viracocha, which means "foam of the sea."

Some years after Viracocha left, it is said, Taguapaca, whom Viracocha had ordered thrown into Lake Titicaca in Collao, returned and started preaching along with others that he was Viracocha. However, although at first they held the people in sway, in the end they were recognized as false and scorned.

This is the ridiculous fable these barbarians have about their creation, and they affirm and believe it as though they had actually seen it happen.[1]

CREATION AND THE HUMAN BODY

The Viracocha cosmogony was recorded in Peru in 1572 by the Spanish soldier Pedro Sarmiento de Gamboa. Similar versions of the same widespread cosmogony can be found in the writings of other chroniclers, Spanish and Andean, of the period. In a few of its recorded versions, at least, this cosmogony displays a Christian influence. Some chroniclers, struck by the similarities between Christianity and Inca religion, were apt to think that an apostle had once proselytized in the region, and interpreted the Viracocha legend as a distorted memory of this apostolic visit.[2] Nonetheless, the cosmogony is one of the earliest existing myths of the Central and Southern Andes[3] and contains fundamental data on Inca cosmology.

In this cosmogony we learn that the human body is modeled after that of Viracocha. The size of Viracocha's body alone is the proper size for humans; the original sculpted giants must be discarded as disproportionate. With its perfect proportions, the human body could now be used as a measure of all other things. Inca measures, in fact, were based on parts of the human body. The smallest measure of length, for example, was based on a finger. The largest was based on the whole body and was the standard measurement of land. Measures of traveling distance were based on the pace, a standard walking length.[4]

Viracocha, as we learned in the previous chapter, was apparently considered by the Incas to encompass both male and female. This would explain how he is able to unilaterally create both sexes. Nonetheless, in the cosmogony, he has a male form. Pachacuti Yamqui, in his version of the myth, describes Viracocha as a bearded man of average height, getting on in years.[5] In one of their temples the Incas had a statue of Viracocha that presented him as "a standing man, his right arm raised, like a person who was commanding."[6]

This characterization of Viracocha as male suggests a privileging of the male body over the female, or at least a closer identification of the former with divine structuring power. This association of masculinity with structuring power was manifested politically in the Inca empire, where most of the positions of power were held by men.[7]

There is a strong indication in the cosmogony, however, that femininity was not originally intended to be subordinate to masculinity. Viracocha creates the Moon, the classic Andean symbol of femininity, with more light than the Sun, the symbol of masculinity. The Sun darkens the Moon out of envy so as to be the brighter of the pair. Translated into human terms, this implies that the Incas, at least theoretically, considered women to be naturally more powerful than men but kept in an inferior position by male stratagems.

The human body, as we saw in the previous chapter, contains and mediates the basic structures of the cosmos within itself. The cosmological basis for this appears to be that the human body is modeled after Viracocha, who encompasses and integrates all dichotomies within himself. Patterned after the body of the creator of the cosmos, the human body is able to serve as a model of the cosmos.

Although humans have the same form as Viracocha, they are apparently not of the same substance. Viracocha and his divinized assistants (who in other versions are his sons) can walk over the sea, humans are tied to the land. One account describes Viracocha as being extremely light (*ligerísimo:* very light and/or fast) and without joints.[8] These qualities reflect Viracocha's spirituality, unbound by physical restrictions, and his essential indivisibility. Along these lines, another account says that the Incas believed that Viracocha was not born of woman and that he was unchanging and eternal.[9] Viracocha's body therefore could be said to represent an original ideal that is not subject to the processes and restrictions governing the human body.

If Viracocha's body is distinguished by its lightness, its ability to walk over the sea, human bodies are characterized by their heaviness, their bond to the Earth. Viracocha gives humans life and form, the Earth gives them their substance. An Andean term for a human being is, in fact, *allpa camasca* (animated earth).[10] Humans are created out of rock and, when their animating force is removed from them, can return to rock, as happened to the first people, who broke Viracocha's precept.

This material foundation of the human being, however, is not completely lifeless or devoid of sacrality, for in Andean thought the Earth itself is sacred and alive. One's material being, therefore, has a spiritual presence of its own that remains with the body after death. For this reason skeletons and mummies were objects of veneration and power for the Andeans. Certain rocks were also worshipped as humans who had reverted to their original substance but who still retained a personal presence. Rock constituted an "other" or double for the human body,[11] a reminder of the earth-bound and ephemeral nature of human life, but also a way for humans to be, within physical limits, unchanging and eternal.

The cosmogony tells us that different peoples are created to inhabit specific places. In the version of the cosmogony given by Bernabé Cobo it is made clear that the original stone models are buried in the earth at different points, from where they later surface as human beings. The sites from which they emerged then became sacred shrines for their descendants.[12] José de Arriaga writes of the Andeans in 1621 that "they also venerate their *pacarinas*, or places of origin. . . . This is one of the reasons why they resist the consolidation of their towns and like to live in such bad and difficult places. I have seen some of them where one can neither go up nor down except on foot; the reason they give for living there is that this is their *pacarina*."[13] Arriaga also notes that this belief in separate places of origin was so entrenched among the Andeans that it was very difficult to convince them of the Christian belief that all peoples descended from a single couple.[14] The Inca prohibition on travel within the empire without special permission[15] built on this Andean tradition in order to facilitate government and the maintenance of a balanced population.

Cobo's version of the cosmogony details as well that the various forms of dress of the different nations were painted on the stone models, thus becoming an integral part of human bodily existence.[16] The Incas regarded their use of clothing as a mark of their identity and civilization. This attitude can still be found in the Andes of today.[17] In fact, a common term for foreigners in the Andes is *qalas*, that is, "naked" or "peeled ones."[18]

Before the Conquest, the basic form of apparel in the Andes was a breechcloth, tunic, cloak, and headband for men, and a dress and shawl for women. The dress of each locality or nation, however, was distinct in its coloring and ornamentation. One of the reasons the inhabitants of the town of Cacha were so alienated by Viracocha is that his dress—described elsewhere as a white gown[19]—was different from theirs. This regional variation in dress was made into law by the Incas so that the place of origin of an individual, and hence kinship loyalties, could be instantly recognized by his appearance.[20]

It is likely that the designs woven into clothes contained condensed information on Andean cosmology.[21] Compare, for example, the pattern Guaman Poma places on the tunic of an Inca (fig. 3) with the drawing Pachacuti Yamqui gives of the cave openings through which the first Incas emerged (see fig. 5). Within this context of elaborated styles and patterns, Viracocha, appearing in a white gown, might have seemed a man without a history or a home. On a metaphorical level, however, Viracocha's white clothes (*llumpacpacha*, with connotations of "pure and whole") signify his transcendence of structural categories.[22]

COSMIC STRUCTURES

Everything has its place and function in the complex, organized system instituted by Viracocha. That human cooperation in upholding this system is necessary is evidenced by the fact that humans are able to disobey the will of Viracocha and destabilize the cosmos. The three instances of violation of the divine plan that occur in the cosmogony all have drastic consequences. In the first two cases the transgressors are reintegrated into the land, either through earth or water, while in the last case fire devastates the land but the transgressors are saved through their repentance. Human conduct, therefore, affects the order of the universe, just as the order of the universe shapes human conduct.

The cosmos created by Viracocha consists of two basic, separately created parts: the earth and the sky. In between the two is the surface of the earth, inhabited by things—people, animals, plants—that emerge from beneath the earth to reach up to the sky. In Quechua these three levels are commonly referred to as Hanan Pacha (Upper World), Cay Pacha (This World), and Hurin or Ucu Pacha (Lower or Interior World). The body, as a model of and for the cosmos, reflects these divisions. The upper part of the body is associated with the structured Hanan Pacha; the middle section,

Figure 3. The Inca Mayta Capac
(From Felipe Guaman Poma de Ayala, *El primer nueva corónica y buen gobierno*, vol. 3, ed. J. Murra and R. Adorno [México: Siglo Veintiuno, 1980], p. 78.)

with its vital organs, with the integrating Cay Pacha; and the lower part with the fluid Hurin Pacha. From another perspective, Hanan Pacha could be understood as the environment outside the body, Cay Pacha as the surface of the body, mediating between exterior and interior, and Hurin or Ucu Pacha as the interior of the body.

In the cosmogony Viracocha acts as a mediating center between opposite halves. The episode where he kneels on the ground and looks up at the sky (a traditional Andean method for swearing an oath),[23] causing fire to come down from above, is one example of this. Another occurs when, in order to call his creation out of the earth, he sends one of his assistants along the coast to his left, the other through the mountains to his right, and takes himself the middle path between them. This mediating role is passed on to humans (and, in Inca cosmology, specifically the Inca), who function as representatives of Viracocha on a more restricted level (as is indicated by Pachacuti Yamqui's diagram in which the human couple is placed under Viracocha and in the center of the cosmos).

In his work Guaman Poma includes an illustraton of the act of Creation according to Christianity that is actually quite Andean in its conceptualization (fig. 4). The Creator is shown standing on the earth with his head in the sky, placing the sun with his right hand and the moon with his left. This illustration, as well as depicting the body as a model of the cosmos, reminds us that Creation takes place through the medium of the body of the Creator, giving the human body, modeled on that of the Creator, a dual role as both created and creator, subject to and manipulator of the cosmic forces.

The periods of Creation outlined in the cosmogony fall within the Andean tradition of dividing time into ages or worlds. According to this tradition, in the middle, and most dramatically at the end, of each age a period of upheaval, known as a pachacuti (world reversal), takes place. The pachacuti is a mediating period of sacred and highly dangerous fluidity during which humans emerge from the natural world and can also return to it, as occurs in the uno pachacuti, world reversal by water, in the cosmogony.

On a microcosmic level a pachacuti can take place within the body of an individual; for example, when a person is struck by lightning, an agent of divinity. Douglas Sharon writes of this in his study of contemporary Andean ritual that "the first charge is believed to kill [the person struck by lightning], the second reduces his body to small pieces and the third reassembles his body."[24] On a macrocosmic level the pachacuti takes place within the body of the cosmos.

Figure 4. The Act of Creation
(From Felipe Guaman Poma de Ayala, *El primer nueva corónica y buen gobierno*, vol. 3, ed. J. Murra and R. Adorno [México: Siglo Veintiuno, 1980], p. 853.)

This is manifested in a myth recorded in the seventeenth century in Huarochiri, Peru, that describes how the world is "turned around" when the divine body (here called Pachakuyuchiq, World Mover) turns around. "They say that when [Pachakuyuchiq] is angry, the earth moves, and also when he turns his face. . . . If he were ever to turn his whole body around, the world would end."[25]

CREATION AND PERCEPTION

The stages of Creation can be compared to those of gestation. The darkness of the first world is the darkness of the womb, the destruction of the first people is the expulsion from the womb, the flood is the fluidity of birth, and the emergence of the new humans from the Earth is the emergence of the child into the light of the outside world. Humans are formed by Viracocha through touch, animated through sound, and then brought out into the light, thereby experiencing the sensations of touch, hearing, and sight in the same order as the Incas very likely conceived of an infant as doing.

While taste and smell are not mentioned in the cosmogony, touch, hearing, and sight are emphasized, indicating the importance of these senses for creation. These latter senses are intimately related in the myth. The human prototypes are sculpted, painted, then named; in other words, given a tactile, visual, and aural identity. Viracocha causes fire to fall from above by kneeling on the earth and looking up at the sky; that is, by being in sensory contact with both halves of the cosmos. Pachacuti Yamqui states that the oral teachings of Viracocha were embodied in the staff he handed on to his followers,[26] a merging of touch and hearing. He describes Viracocha's burning of the mountain as a "terrifying miracle, never before heard in the world,"[27] interrelating a visual manifestation with oral tradition.

Of these three senses, hearing plays the most important role in the cosmogony. Human beings are called into life and purpose by the divine word. The creative primacy and potency of the word of Viracocha is manifested in an Inca prayer recorded by Cristóbal de Molina "of Cuzco" in 1574, which states, "O Viracocha, who . . . said let this be a man and let this be a woman, and by so saying made, formed, and gave them being."[28]

More than anything else, Viracocha is the one who *speaks.* Pedro de Cieza de León writes in his version of the cosmogony that Viracocha "spoke to [his created peoples] lovingly and gently, admonishing them to be good and not harm each other, but rather love each other."[29] Molina of Cuzco describes how Viracocha named all the plants and animals and instructed the people in their uses.[30] Pachacuti Yamqui says that Viracocha "spoke all languages better than the natives," and has a huaca refer to the Creator as a "chatterbox."[31]

Viracocha's words are meaning coupled with the dynamic force of sound. Cobo writes that "with his word alone [Viracocha] made the corn and vegetables grow."[32] In the Huarochiri myth the words of the Creator (Pachacamac, Animator of the World) are so powerful that they can shatter the whole cosmic order. "I am the one who shakes you [the Inca] and the whole world. I have not spoken because I would not only exterminate your enemies, but I might destroy all of you and the whole world at the same time. That is why I have remained in silence.[33] Perhaps this is the answer to the ardent desire expressed in many Inca prayers to be spoken to by the Creator.

When the Creator speaks, the world responds. As representatives of the Creator, humans participate in the creative power of speech. We are told in the cosmogony recorded by Cobo that Viracocha "gave each nation the language they are to speak and the songs they are to sing."[34] In this way humans are granted the gift of language and the ability to enter into dialogue with each other and their Creator. In Inca cosmology it is the Inca who, like Viracocha, is endowed with the ability to animate and order the world with his voice. This belief survives in modern Andean traditions in which the Inca is the one who moves rocks with his voice and makes mountains speak.[35]

The Incas' own name for their language, Quechua, was Runa Simi (Man's Mouth), which reminds us that the words that order the world are imbued with the physical presence of the speaker, they are carried in sound that originates inside the body and travels outward. To speak is to give birth, to create, and to make a commitment of participation.

The actions of hearing and speaking can be compared to the actions of inhaling and exhaling; the exchange of essence and power with the outside world. Indeed, in one Inca prayer Viracocha's act of creation by speech is associated with the act of breathing a spirit into something, signified by the Quechua verb *çamay*, "to animate."[36] If speech is life, silence is death. The name given by the Andeans to the place of the dead was Upamarca, "Land of Deaf-Mutes."

The first order given to the humans called out of the Earth in the cosmogony is to "hear and obey" the commandment of Viracocha. The transgressions listed result from the refusal to hear and obey Viracocha's word. Pachacuti Yamqui writes that "the Indians of that time used to laugh at [Viracocha], saying, what a talkative man, and although he always preached to them they didn't listen."[37]

This refusal to hear can be likened to a refusal to be fertilized and to enter into a productive state of culture. By denying Viracocha's word, humans remain in a barren, uncivilized condition. This condition is denoted in Quechua by the word *purum*, which means savage, unconquered, barren, and virgin. The term *purum soncoyoc* (savage heart) expresses the state of those who refuse to attend to the divine word. According to González Holguín it means, "he who does not try or wish to know good things about his salvation . . . nor lets himself be cultivated or helped to produce fruit."[38] Refusing to hear, nonetheless, is the correct response to inauthentic words. In the cosmogony Viracocha warns his people that false Viracochas will appear among them and that the people should not believe what they say.

The first period of the world, generally referred to as one of darkness, is called Purun Pacha, (savage world), by Pachacuti Yamqui. He describes this period as plagued by *hapiñuños*, evil beings with long grasping breasts, and characterized by constant warfare.[39] Purun Pacha, with its associations with virginity, its darkness, its chaos, and its long-breasted hapiñuños, has feminine connotations. The second period, in contrast, is one of light, cultivation, and order—masculine characteristics. Pachacuti Yamqui tells us that by this period the hapiñuños have been banished, crying out, "We are defeated," and describes how Viracocha angrily destroys an idol in female form.[40] (Significantly, Viracocha uses water, a female element, to punish the first humans, and fire, a male element, to punish those of the second creation in the cosmogony.) The chaos of the first age was thus possibly symbolically due to its excessive femaleness. When the male sky is created to counterbalance the female Earth this problem is resolved.

To a certain extent Viracocha, who creates the sun and the moon in order to achieve a more perfect world, is associated with light, and therefore with vision, as well as with sound and hearing. Pachacuti Yamqui, perhaps influenced by Platonic ideals, refers to Viracocha as the "true Sun," while the modern scholar Zuidema writes that the route Viracocha takes reflects the path of the sun across the sky.[41]

It is only after light, and thereby sight, has been created by Viracocha that the world is structured. In Andean tradition, for example, stones were able to move in the first, dark age, but with the creation of the sun they became fixed in their place.[42] Cobo writes that Viracocha's primary task was "ordering and arranging all of creation, putting each thing in its proper place."[43] Sight must likewise fix objects in their place, detain them in space, in order to perceive them well.

However, while light and sight may allow for a more structured cosmos than sound and hearing, there is no doubt that the former are dependent on the latter in Andean cosmology. In the cosmogony humans must first hear before they are able to see. According to Cieza, Viracocha cured the blind with his word alone, thereby giving them sight through their faculty of hearing.[44] The sun and the moon arise from Lake Titicaca at Viracocha's command. Conversely, in the Huarochiri myth the Creator, Pachacamac, tells the "light of the sun," the Inca, that he can destroy him with his word alone.[45]

If Viracocha's structuring role and creation of light associate him with sight, his fluid qualities associate him with the fluidity of sound. Viracocha's journey down to the sea, over which he disappears, is like that of a river, flowing down to the sea. In some myths Viracocha is also said to have arisen out of water.[46] His name, Viracocha, can be interpreted to mean "sea foam." Viracocha's very movement, light and smooth (without joints), is fluidlike. This fluidity is in keeping with the orality that characterizes Viracocha.

Finally, the cosmogony itself was communicated orally by the Andeans. For literate peoples, accustomed to sacred traditions existing in definitive textual versions, the variations and contradictions among the Andean cosmogonies might seem evidence of their unreliability. For the Andeans, however, the fluidity of the myth testified to its dynamism and life, and to the continued creative activity of the sacred word in the world.

4
MYTHS OF THE INCAS

THE ORIGIN OF THE INCAS

In the old times people lived like wild beasts, without religion or laws, houses or towns, without cultivating the earth or wearing clothes, and without having separate wives. Our father the Sun, seeing people in this state, took pity on them. He sent Manco Capac and Mama Ocllo Huaco, his son and daughter by the Moon, to teach the people to worship their father the Sun, and to give them laws to live by so that they might live in houses and towns, and cultivate the earth and raise animals like rational and civilized men.

The Sun placed his two children in Lake Titicaca and gave them a golden staff, bidding them to set up their court at the site where it would sink into the ground at one thrust. Manco Capac and Mama Ocllo Huaco left the lake and walked north. They spent the night in a dwelling the Inca Manco Capac called Pacarec Tampu, House of the Dawn, because he left it as the sun rose. When they reached Huanacauri hill in the valley of Cuzco the staff sank in the ground with a single thrust, disappearing into the earth, whereupon they determined to establish their dwelling in the valley.

Manco Capac then went north and Mama Ocllo Huaco south, telling the people they met of their divine mission and calling them out of the wilderness. The people recognized them as children of the Sun by their clothes, their pierced ears, their words, and their faces, and followed them back to the valley. There they founded the city of Cuzco. Those people convoked by Manco Capac formed Hanan, or Upper, Cuzco, and those convoked by Mama Ocllo Huaco formed Hurin or Lower, Cuzco. Manco Capac ordered that the upper half should have precedence over the lower half, as elder brothers over younger brothers, or right and left arms, as the former were gathered by a man and the latter by a woman. The same division was made in all the towns, large and small, of the empire.

The Inca Manco Capac taught the men agriculture and how to build irrigation canals, and the Coya (queen) Mama Ocllo Huaco taught the women how to spin and weave and make clothes, and all other domestic duties. In short, everything pertaining to human life was taught them, with the Inca serving as the teacher of the men and the Coya as the teacher of the women.[1]

El Inca Garcilaso de la Vega, the half-Inca, half-Spanish chronicler, recorded the Inca myth condensed here in 1609. In it we learn that before the arrival of the Incas the people lived in a state of savagery, virtually indistinguishable from nature. The Incas come from the sky, and therefore from structure, to bring the peoples of the Earth out of the chaos of their natural condition into a state of divinely sanctioned order. To do so, they must teach the people to conduct themselves correctly both with respect to their bodies and to nature; specifically, to practice agriculture, to live in towns (i.e., to have a settled communal existence), to wear clothes, and to regulate their sexuality. A civilized person is thus one who transcends Earth and nature and participates in the structure of the sky. A savage is one who is identified with the Earth and nature and therefore unable to interact with both levels of the cosmos or serve as a mediator between them.

The golden staff given to Manco Capac and Mama Ocllo Huaco symbolizes the divine power of the sky that fertilizes the earth at a receptive point. The staff sinks into Huanacauri hill, connecting the three levels of the cosmos and establishing Cuzco as a chosen site of communication between the sky and the Earth. The *axis mundi* thereby established is composed of three overlapping axes—the golden staff of the sky, the mountain of the earth, and the body of the Inca. This union is accomplished through the medium of the Inca, who is thenceforth the preeminent mediator of cosmic exchange and order.

José Arguedas recorded a modern legend that displays similarities with this myth. The *wachok* (fornicators), who belong to an age before that of the savages, enter into the heart of the mountain through its veins of water to find the water's source. In order to do so they wear golden drums on their heads and golden and silver clothes. When they come out they give the people fertile lands.[2] This legend seems to refer to the union of the sky—represented by the gold and silver clothes—with the Earth, accomplished through the medium of the bodies of the mythical fornicators. The result of this mediated union is fertility, a dynamic and productive cosmos.

In the myth of the origin of the Incas we see the marked male/female dualism characteristic of Inca culture. When Manco Capac and Mama Ocllo Huaco convoke the people from the wilderness, the former travels ahead (north), in the symbolic direction of male structure, and the latter back (south), toward Lake Titicaca and female fluidity. Those people gathered by Manco Capac form the dominant Hanan Cuzco, associated with male, sky, and structure, while those gathered by Mama Ocllo Huaco form the subordinate Hurin Cuzco, associated with female, earth, and fluidity. Manco Capac teaches the men the traditionally male occupation of agriculture, and Mama Ocllo Huaco teaches the women traditionally feminine domestic duties, such as spinning.

While Mama Occlo Huaco symbolizes the feminine, it is definitely the *structured* feminine that she symbolizes, the civilized Inca woman who knows how to spin and weave. She descends from the ultimate Inca symbol of structured femininity, the moon, with its controlled cycles and its illumination of the dark night.

Other versions of the myth of the origin of the Incas allow women a more aggressive role than that generally allotted them in the Inca cosmological-political structure. Pedro Sarmiento de Gamboa, for example, describes "Mama Guaco," Manco Capac's sister, as extremely strong and fierce. She throws the golden staff that fixes in the ground at Cuzco, thereby assuming the role of cosmic integrator.[3] This association of femininity with force seems to be more in keeping with popular Andean tradition, in which the female Earth plays a dominant role, than with Inca cosmology, in which femininity is closely linked with conquered elements and peoples.

As well as being associated with male and female, Upper and Lower Cuzco are compared to an elder brother and a younger brother by Garcilaso de la Vega. While the male/female division of the city is an expression of complementary opposition, the elder brother/younger brother division provides a model for antagonistic similarity, indicating that the two halves may sometimes vie with each other for power.[4] A ritual battle, known in Quechua as *tincuy* (encounter), in fact took place annually between the two moieties of the city.[5] However, the analogy of the two halves of Cuzco to the right and left sides of a body emphasizes that Cuzco, and all other communities participating in its structure, is fundamentally one.

In Garcilaso's myth the Incas have three symbolic birthplaces: the sky (they are born of the Sun and the Moon); water (they emerge from Lake Titicaca); and the Earth (they come out of the dwelling Pacarec Tampu). This enables them to represent and integrate all three basic elements.

Other, perhaps earlier, versions of the myth stress the relation of the Incas to the Earth. Pachacuti Yamqui, for instance, gives the Incas human parents, represented in symbolic form as two trees (fig. 5).[6] Sarmiento de Gamboa, relating a popular version of the

Figure 5. The Three Windows from Which the First Incas Emerged, and the Two Trees Representing Manco Capac's Parents
(From Joan de Santacruz Pachacuti Yamqui, "Relación de antigüedades deste reyno del Pirú," in *Tres relaciones de antigüedades peruanas,* ed. M. Jiménez de la Espada [Asunción, Paraguay: Editorial Guaraní, 1950], p. 217.)

myth, names Pacaritambo (Pacarec Tampu) as the birthplace of the Incas.[7] He describes it as three windows in the side of a hill. The Incas—four brothers (Manco Capac, Ayar Auca, Ayar Cache, and Ayar Ucho) and their sisters-wives, emerge from the middle window. They decide to conquer the land, but Ayar Cache threatens the enterprise with his violent and disorderly conduct. After an argument Manco Capac sends Ayar Cache back to Pacaritambo to fetch something and has him walled in.

When the remaining Incas reach Huanacauri hill Ayar Ucho turns into a stone huaca on the top. He tells his brothers and sisters to remember him in all their ceremonies, particularly that of *huarochico,* the puberty rite of boys. Ayar Auca then turns into a stone post of possession in between two rivers in the valley below, on the future site of the Inca temple, Coricancha. After this, Manco Capac and Mama Guaco go on to conquer the surrounding peoples and found Cuzco.[8]

This version of the myth exhibits the fluidity between body and land characteristic of the pachacuti, the destructured intermediate period between ages. The Incas emerge from the Earth and three of them return to the Earth. Ayar Cache, associated with chaos, is assimilated into Ucu Pacha, the chaotic Lower World. Ayar Ucho, associated with sacred structure, is transformed into a huaca on top of a hill, representing thus the structure of Hanan Pacha, the Upper World. Finally, Ayar Auca becomes a stone post of possession on the site of the Inca temple in the valley, symbolizing the worldly order of the Incas and the mediating role of Cay Pacha, This World. The fact that the word *auca* can mean "adversary" also implies the need to maintain the Upper and Lower worlds distinct and to protect the center from being destabilized by the forces outside.

The centrality of the Inca order in the cosmos is indicated at the start of the myth when the Incas emerge from the middle of the three windows of Pacaritambo. Garcilaso de la Vega relates that the Incas had the middle window of this shrine decorated with gold and precious stones. He then goes on to give the meaning of Cuzco as "navel," thus reaffirming Inca centrality with a bodily metaphor.[9]

The myths of the origins of the Incas show parallels with the Viracocha cosmogony and can perhaps be viewed as variations on this myth. In Garcilaso's myth the Incas appear at Lake Titicaca (the center of origin), as does Viracocha, and take the same route northward as he does. The pre-Inca age of savagery in the myth corresponds to the initial period of darkness in the cosmogony. The Incas are sent by the Sun to call the people out of the wilderness into civilization. Viracocha creates the Sun and then calls people out of the Earth and gives them culture. Like Viracocha, the Incas serve as models for their subjects. It is possible, in fact, that the word *inca* signified the original model of every being, as well as ruler.[10]

Viracocha's three helpers are similar to Manco Capac's three brothers in Sarmiento's myth. Taguapaca, the helper who disobeys the orders of Viracocha and thereby embodies antistructure, is sent back into the fluidity of the Earth, as is Ayar Cache, the brother who disobeys Manco Capac. Both Viracocha and the Incas carry staves that help them order and animate the world. Pachacuti Yamqui, in fact, states that the staff of the Incas was that left behind by Viracocha. It is with this staff, according to Pachacuti Yamqui, that Manco Capac is able to prevent himself from being turned into stone by a huaca.[11] (A staff, called *tupa yauri*, royal scepter, in fact formed an essential part of the royal insignia of the Incas.)

While the Incas have a similar structuring role as Viracocha, they are nevertheless part of the created world and partake of its nature. Viracocha contains both male and female within himself and thus needs no partner (Molina of Cuzco explicitly states that Viracocha did not have a wife).[12] The closest the Incas, as participants in the dualistic cosmos, can come to this ideal is to form a male-female pair in which they are united both by marriage and by blood (brother-sister).

As Viracocha ultimately transcends worldly dualism and structures, the inhabitants of the Earth find him alien and turn against him. As the Incas, however, manifest the highest order of worldly dualism and structure, the peoples of the Earth, according to Garcilaso, recognize them as children of the Sun, the preeminent symbol of structure, and follow them. This distinction is expressed in part by the different clothes worn by Viracocha and the Incas, which contribute to the rejection of the former and the acceptance of the latter. Whereas Viracocha wears a white gown, placing him outside worldly categories, the gowns of the Incas, according to Cieza de León, are richly decorated,[13] placing them at the head of worldly categories.

This distinction is perhaps further manifested by the contrast between the plain oval (unity) that represents the Creator and the elaborate rectangles (diversity) that represent the created Incas (see figs. 2 and 5) according to Pachacuti Yamqui. Finally, while Viracocha disappears over the sea and leaves behind the structured world, the Incas remain in Cuzco, the earthly center and model of structure, to serve as mediators of power and order.

THE REVERSAL OF THE WORLD

The traditional sequence of Inca rulers from Manco Capac to the Conquest, with possible dates of reign for the later rulers, is:

Manco Capac	
Sinchi Roca	
Lloque Yupanqui	
Mayta Capac	
Capac Yupanqui	
Inca Roca	
Yahuar Huacac	
Viracocha Inca	
Pachacuti	(1438–1471)
Tupac Inca	(1471–1493)
Huayna Capac	(1493–1525)
Huascar	(1525–1532)
Atahualpa	(1532–1533)[14]

It was with the ninth Inca, Pachacuti, that the area controlled by the Incas began to reach imperial proportions, according to Inca tradition. The new era was inaugurated by Pachacuti's victory over the rebellious Chancas (a central Andean people), who were on the verge of capturing Cuzco and overthrowing the Inca state. The traditions dealing with this crucial event are not as markedly mythological in character as the previous myths we have examined; however, they manifest a symbolic structure consistent with that of the earlier myths.

Following is an abbreviation of the tradition of the battle against the Chancas as recorded by Juan de Betanzos in 1551.

> Viracocha Inca received his name after having been spoken to by Viracocha one night. During his reign Cuzco was threatened by the neighboring Chancas. Viracocha Inca, finding his forces outnumbered by those of the Chancas, retired to a nearby mountain with his son and heir, Inca Urco, and the inhabitants of Cuzco. The youngest of Viracocha Inca's sons, Yupanqui (later called Pachacuti), stayed behind in the city with three companions.
>
> Yupanqui sent a message to his father saying that Cuzco should be defended. His father responded that he was a man who spoke with God and that God had told him that he could not resist the Chanca forces. In a state of desolation, Yupanqui went outside the city and prayed to Viracocha, the Creator, for help. While he was sleeping, Viracocha appeared to him in the form of a man and told him he would send troops to help him attain victory.
>
> The day of the battle with the Chancas a great number of unknown people appeared, and with their aid Yupanqui was able to defeat the Chancas. After the battle these unknown people sent by Viracocha disappeared. Despite this victory, Viracocha Inca refused to recognize Yupanqui as his heir.
>
> Yupanqui, now in command of Cuzco, examined the city and found it to be built on swampy ground with small, poor houses distributed at random. The first thing he did was build a temple to the one he had seen in the vision before the battle. Yupanqui had called the figure in the vision Viracocha, but now he reasoned that, as he had never actually been told his name, and as the vision had shone so brightly that he seemed to see its light before him all day, the figure must, in fact, have been the Sun. He therefore dedicated the temple to the Sun and established the rites that would be celebrated in honor of the Sun. As the vision had addressed Yupanqui as son, the people thereafter referred to Yupanqui as the son of the Sun.
>
> Yupanqui ordered that the land should be divided up among the inhabitants of the area, with the boundaries between one property and the next clearly delineated. He had food and clothes stored in great quantities in the city for times of need. He rebuilt the irrigation canals and fortified the banks of the two rivers running through Cuzco, which regularly flooded the city. He arranged for people from neighboring communities to intermarry so that they would live together in peace

and friendship. He divided the year into months and set up posts to serve as sundials so that time could be measured.

Yupanqui made a model in earth of the new Cuzco he wanted to build. He then demolished the city and rebuilt it according to his plan. That part of the city below the temple of the Sun, he called Hurin Cuzco, and decreed that his allies and Incas of mixed blood should live there, while pure-blooded Incas should occupy that part of the city above the temple of the Sun, which he called Hanan Cuzco. The whole city together he named Lion's Body, saying that the inhabitants were the lion's members and that he was its head.

The citizens of Cuzco then decided that Yupanqui should be officially inaugurated as the Inca. When they went to discuss it with him they found him drawing pictures of bridges and roads. As this was foreign to their understanding, they asked Yupanqui what he was drawing. He, in turn, inquired of them what they had come to ask him. On learning of their mission, Yupanqui refused to be inaugurated as Inca without the approval of his father.

Viracocha Inca came to Cuzco at the request of its citizens, and Yupanqui met him and addressed him as his lord and father. On seeing how well ordered the city was, and the great esteem in which Yupanqui was held by everyone, Viracocha Inca said to his son, "You are truly the son of the Sun." He then took the royal fringe off his head and placed it on his son's, renaming him Pachacuti, World Reverser.

Pachacuti then ordered Viracocha Inca to drink *chicha* [corn beer] out of a dirty cup, telling his father that, as he had acted as a woman, he was one, and that he should drink only out of dirty cups. Viracocha Inca, prostrate before Pachacuti, responded, "My cruel father! I know my sin." After that Pachacuti treated Viracocha Inca with all honors. When the festivities were over Viracocha Inca left Cuzco once again, and Pachacuti went on to conquer many peoples and subject them to the dominion of Cuzco.[15]

This tradition has parallels with both the Viracocha cosmogony and the myth of the origin of the Incas. Viracocha Inca (the title *Inca* could come either before or after a name) has obvious associations with Viracocha. Betanzos says that Viracocha Inca received his name after having been spoken to by the god. The chronicler Cieza de León says that the Inca was called Viracocha because he came from foreign parts and had a foreign style of dress and appearance, as did the Creator.[16] He describes how Viracocha Inca set a rebellious town on fire, as did Viracocha, and has him returning from a visit to the Lake Titicaca region prior to the battle with the Chancas, thus making him travel the same route as Viracocha.[17] Viracocha Inca's abandonment of Cuzco can be compared to Viracocha's abandonment of the created world.

Cuzco, which represents the world in miniature, thus abandoned, is on the point of succumbing to the forces of chaos. This is the unstructured intermediate period between ages that we find between the age of darkness and the age of light in the cosmogony and between the age of savagery and the age of the Incas in the myth of the origin of the Incas. The behavior that Cieza de León attributes to Inca Urco, Viracocha Inca's heir, associates him with chaos and fluidity. He does not build houses, he has affairs with married women, he drinks to excess, and vomits and urinates without shame in the open.[18]

According to Betanzos, when Yupanqui (hereafter referred to as Pachacuti) examines the city he finds it in a state of disorder, built on swampy ground, subject to floods (uncontrolled fluidity), and with poorly built houses distributed at random (chaos). Another manifestation of the fluidity between land and body characteristic of the period between ages is found in the Inca tradition that the unknown warriors sent to assist the Inca prince were stones that changed into people for the battle and then back into stones again.[19] The event that makes it possible for order to be restored in the Inca world is the defeat of the Chancas, who represent the forces of chaos.

With the defeat of savagery, Pachacuti becomes the new Manco Capac, the founder of the Inca order. Thus, like Manco Capac, he takes on the title of son of the Sun. Sarmiento de Gamboa mentions that Pachacuti visited and entered inside the shrine of Pacaritambo through which the first Incas had emerged.[20] Through this action Pachacuti becomes symbolically identified with Manco Capac.

Manco Capac and Pachacuti were, in fact, the two fundamental figures of Inca imperial tradition. All Inca laws were apparently attributed to one or the other.[21] The prominence of these two figures in Inca mythology explains why the Inca leader who rebelled against Spanish rule after the Conquest took the name Manco Capac Pachacuti Yupanqui.[22] In doing so he was hoping, no doubt, to associate himself with the world-reversing power of these culture heroes.

As the myth of the origin of the Incas mirrors the Viracocha cosmogony, so does the account of the battle with the Chancas reflect both these traditions. Pachacuti Yamqui writes that Pachacuti is able to turn the situation around in his favor during the battle when he takes up the staff that was originally Viracocha's and then Manco Capac's.[23] Pachacuti's three companions are reminiscent of Viracocha's three assistants and Manco Capac's three brothers. In the

Viracocha cosmogony the age of darkness is followed by the age of light after Viracocha creates the sun. In the myth of the origin of the Incas the age of savagery is transformed into an age of culture when the Incas are sent to Earth by the Sun. Pachacuti is able to defeat the forces of chaos after he has a vision of what he decides is the Sun, and he inaugurates the new period of order by building a temple to the Sun. Just as fluidity, therefore, indicates a breakdown of categories in Inca mythology, light signals the establishment of order.

When Pachacuti makes a model in earth of the new Cuzco he is repeating not only Manco Capac's original creation of the city, but also Viracocha's modeling of humans out of stone. This analogy is made clearer when Pachacuti uses the metaphor of a lion's body to refer to the city. Whereas Viracocha creates the human body, therefore, Pachacuti creates the social body. Pachacuti himself is associated with the upper half of this body (Hanan Cuzco) and with the temple of the Sun that mediates between the two halves.

Molina of Cuzco reports that, on his way to his father, Pachacuti had a vision of the Sun as a man with a lion's head between his legs, another lion over his shoulder, and a serpent running up his back.[24] If this divine figure can be taken to represent the cosmic body, the two lions would symbolize the separate structures of Earth and sky, and the serpent the integrating axis that connects the two halves of the cosmos. Sarmiento de Gamboa mentions that the lightning deity, Illapa, gave Pachacuti a two-headed serpent as a talisman,[25] which emphasizes Pachacuti's similar role as cosmic mediator.

For Pachacuti and Manco Capac, Cuzco serves as a dualistic microcosmos through which humans are structured, just as humans were originally structured by Viracocha after the creation of the dualistic cosmos. Pachacuti, however, takes his structuring activity further than either Manco Capac or Viracocha. He creates storehouses of food and clothes, thus bringing agriculture to its cultural apex, as depicted in Pachacuti Yamqui's diagram of Inca cosmology (see figs. 1 and 2). He intermarries people from different communities to promote social harmony, thereby applying the complementary principle of karihuarmi (man-woman) to social bodies.

Cobo writes of Pachacuti that "nothing was overlooked by him in which he did not impose all good order and harmony."[26] This emphasis on structuring is graphically evidenced by Pachacuti's division of the land into separate properties with clearly marked boundaries. Pachacuti's rebuilding of irrigation canals and fortification of riverbanks manifests his desire to keep categories integrated, not flooded or confused. By dividing the year into months and creating a system to measure time, Pachacuti structures time as well as space, thus establishing a total order.

While Pachacuti can be seen as repeating and continuing Viracocha's activities, he can also be seen as opposing and supplanting Viracocha. Pachacuti takes over Cuzco, the microcosmos, after it has been abandoned by Viracocha Inca (who symbolizes the Creator). The vision that Pachacuti at first attributes to Viracocha the Creator, he later attributes to the Sun, perhaps to distinguish himself from his father, who communicates with Viracocha. Molina of Cuzco reports that Pachacuti saw the vision of the Sun while going to see his father. This indicates that the Sun has supplanted Viracocha Inca as Pachacuti's father. Viracocha Inca is brought back to Cuzco, the created world, to recognize Pachacuti's superior structuring power as son of the Sun, and to confer on him the name Pachacuti, The One Who Reverses the World.

Pachacuti forces his prostrate father to drink dirty chicha and calls him a woman, associating him thus with the Lower World, disorder, fluidity, and femininity. When Viracocha Inca enters the city Pachacuti addresses him as father. After the rite of humiliation described above, however, Viracocha Inca addresses Pachacuti as father. With the reversal of order, the son becomes the father of the father, and the created Sun takes primacy over Viracocha, the Creator.

Sarmiento de Gamboa mentions that Pachacuti placed a figure of Viracocha to the right of the image of the Sun in the temple, and a figure of Illapa, Thunder-Lightning, to the left, adding that Pachacuti took the statue of Illapa to be his *huaque* (symbolic brother).[27] These three divinities can be said to form a triumvirate in Inca religion. In his analysis of Inca mythology Zuidema finds Viracocha to be associated with the rainy season, night, and fertility; Illapa with war and liminal periods; and the Sun with the dry season, day, and political administration.[28] During the liminal period of the battle with the Chancas, Pachacuti can be seen as identified with his brother Illapa, the deity of transition, while in the period of reconstruction afterward he becomes identified with his father the Sun, the deity of structure.

Some chronicles show an attempt to overcome the opposition between Viracocha/Viracocha Inca and the Sun/Pachacuti. Pachacuti Yamqui does this by identifying Viracocha as the "true Sun," of which the worldly sun is only a representative.[29] Garcilaso de la Vega attributes the victory over the Chancas to Viracocha Inca.[30]

This may reflect the interests of factions within Inca society with closer ties to Viracocha Inca than to Pachacuti, or simply a desire to give Viracocha Inca a more respectable role in the proceedings in keeping with his sacred associations.

Garcilaso de la Vega describes Viracocha Inca as having a vision of a man holding an animal on a leash. This man tells Viracocha Inca he comes from his father the Sun and promises him divine aid in his endeavors on behalf of the empire.[31] The animal held on a leash by the divine figure appears to be symbolic of the barbaric nations the Inca is to subjugate and civilize.

A few of the nations subjugated by the Incas, however, were considered by the Incas to be too closely allied to the Earth to be fully civilized. Garcilaso writes of a group of these conquered by Pachacuti's son, Tupac. "They lived like wild beasts scattered about the countryside, which made it more difficult to instruct them and bring them into a state of civilization than to simply conquer them. [The Incas] taught them to make clothes and footwear and to cultivate the earth by making irrigation canals and terraces . . . but the Incas did not build temples for the sun nor houses for the chosen virgins because of the unfitness and vileness of the people."[32]

The fate of those subjects who refused Inca rule and civilization, and thereby, according to Inca ideology, refused to realize their full human potential through participation in the Inca order, was to be cut off from the sacred structure of the sky and integrated into the chaotic Earth. In this regard, Garcilaso notes that the captives of the battle with the Chancas were warned to either be good vassals or be "swallowed up alive" by the Earth.[33]

The Inca ideological program, therefore, was to tranform chaotic, savage bodies into structured, civilized bodies. This program was part of the basic Incâ cosmological cycle, evidenced in the Viracocha cosmogony and in the myths presented here, in which chaos is structured, then degenerates into chaos again, and is restructured.

In the myth of the battle with the Chancas, Cuzco, originally structured by Manco Capac, has fallen prey to the powers of chaos, symbolized by the savage Chancas. The reigning Inca has effectively deposed himself by abandoning the city. The Inca who is the heir to the throne, Urco, and who should serve as a symbol of structure (*urco,* in fact, means "male," and is therefore associated with structure), embodies the structural decay of the state by his dissolute behavior. Pachacuti reverses the situation by defeating the Chancas and supplanting the reigning Inca and his heir. Sarmiento de Gamboa states, in fact, that Pachacuti kills Inca Urco after knocking him into a river.[34] Through this action Pachacuti unites Urco with the fluidity he symbolized and integrates him into the Earth (as did Viracocha with Taguapaca and Manco Capac with Ayar Cache).

Pachacuti is then able to restructure Cuzco, fortifying riverbanks, marking boundaries, reaffirming the division of the city into two halves, and establishing a mediating center. With its structured fluidity and complementary dualism, Cuzco can now serve as an ideal both for the empire and for the human body.

THE INCA SENSORY ORDER

The myths of the Incas contain few references to taste and smell. One notable exception is found in the myth of the origin of the Incas where the names of two of Manco Capac's brothers, Cache and Ucho, refer respectively to salt and red pepper, the two most important Andean condiments. (The meaning of the word *ayar*, which precedes the names of Manco Capac's brothers, is not known.)

Garcilaso de la Vega writes of the allegorical use of the names of these condiments that "by salt . . . the Andeans understand the Inca's teaching on the natural life, and by pepper the pleasure it gave them."[35] This interpretation is difficult to reconcile with Cache's (salt's) associations with the chaotic Lower World and Ucho's (pepper's) association with the structured Upper World; however, it does indicate that these condiments were symbolically meaningful for the Incas.

It is probable that salt is mythologically connected with the Lower World (Cache is sealed in a cave) in part because it is found on or under the ground, while pepper is associated with the Upper World (Ucho becomes a huaca on a hill) because it grows above the ground. Interestingly, one possible interpretation of Manco Capac is "noble cereal,"[36] which then suggests the image of a neutral food in between the opposites of salt and pepper.

According to modern Peruvian Andean tradition, the inhabitants of the first age of the world did not use either salt or pepper.[37] In the myth of the origin of the Incas the appearance of Cache and Ucho seem to represent the introduction of these two condiments into the world. The use of salt and pepper would consequently form an important part of the culture the Incas impart to their subjects in the myth.[38]

It was not only the use of salt and pepper that was culturally important to the Incas, however, but also the abstention from their use at certain times. Abstaining from salt and pepper, in fact, was the primary element of the ritual fasts that were an essential prelude to most Inca ceremonies. This ritual abstention is perhaps symbolized in the myth by the walling up of salt in the cave, and the petrification of pepper.

Touch plays a greater role in Inca mythology than either taste or smell. The most prominent use of touch in Inca myths is to accomplish or prevent a transformation: Ayar Ucho turns to stone after being touched by a stone huaca;[39] Pachacuti models the new Cuzco in clay before having it built; Viracocha Inca passes on the kingship to Pachacuti by placing the royal fringe on his head, and so on. On the other hand, Viracocha uses his staff to put out the fire that threatens to destroy the world, and Manco Capac hits a huaca with his staff to prevent himself from being turned into stone.[40] This use of touch in Inca mythology emphasizes its power to either join or separate, attract or repel.

As in the creation myth, in the myths of the Incas touch is often closely related to hearing and sight. The sacred staff that appears in the different myths provides a good example of this sensory overlapping. In the previous chapter we learned that Viracocha's staff contained his teachings and thus represented a fusion of touch and sound. According to one chronicle, when Manco Capac uses this staff to strike a huaca that has captured his brother and sister the huaca responds: "If you hadn't brought that staff that old chatterbox [Viracocha] left you, . . . I would have dealt with you according to my will too."[41] On being hit by the staff, the huaca is symbolically addressed by Viracocha's word, which the staff embodies, and is thereby obliged to respond.

Later in the work, when the Chancas are attacking Cuzco, Pachacuti hears a voice from the sky telling him to take the staff from the temple. With the staff, and the divine words contained in it, he is able not only to "animate his soldiers and captains so they will fight," but to animate rocks so that they will become soldiers.[42]

The power of light and sight is merged into the staff in the myth of the origin of the Incas, in which the staff symbolizes a ray of sun. Pachacuti Yamqui states that the staff containing Viracocha's teachings turned into gold, the metal of the sun, on the birth of Manco Capac.[43] This suggests an evolution from touch to hearing to sight, such as occurs in the cosmogony, in which humans are molded by touch, animated by sound, then brought into the light.

The Incas are strongly identified with light in their myths. The foremost example of this, of course, is the Incas' reputed descent from the Sun. One myth goes so far as to state that Manco Capac was engendered by a ray of light entering a window,[44] thereby substituting light for touch as the medium of procreation. The Incas' emergence from Pacaritambo, House of the Dawn, as the sun rises, symbolically equates their birth with the birth of the sun. The sun, in fact, was believed to travel underneath the earth at night.[45] The Incas thus emerge from the earth at dawn, as the sun does.

The Sun was a very useful patron god for the Incas, as he could be perceived, and therefore worshipped, from anywhere in the empire. In order for the Inca's subjects to fully appreciate the sacred role of the Sun in Inca cosmology, however, it was essential that they properly value the sense of sight. When the Incas bade their subjects to turn from the worship of the "dirty and ugly" creatures of the Earth to that of the "resplendent and beautiful" Sun,[46] they were also asking for a change from multisensory sacred models to a sacred model that was primarily visual.

It is significant that it is the *light* of the sun, its visual splendor, that is emphasized in Inca cosmology, and not its heat. This is in contrast to the indigenous cosmologies of Central America, where the reverse is true.[47] The Inca emphasis on light over heat is also manifested in Quechua, in which there are far more terms dealing with the visual qualities of light than with qualities of heat (see the appendix).

The following Inca myth describes how the Incas, as sons of the Sun, are able to dominate the orality of the huacas with their sight.

> The Inca Capac Yupanqui wanted to see how the huacas spoke with their followers. The ministers of a huaca took him into a dark hut and called on their huaca to speak to them. The spirit of the huaca entered with the sound of the wind, leaving everyone chilled and afraid. The Inca then ordered that the door and windows be opened so that he could see the huaca. When the door was opened the huaca hid his face. The Inca asked him why, if he was so powerful, was he afraid to raise his eyes? The figure, which was of frightening size, very ugly, with thick, curly hair and a foul odor, shouted like thunder and rushed out.[48]

In this myth the Inca, through the power of sight, defeats the local huaca, who communicates through sound and who is evidently a thunder deity. Light reveals the huaca, who was previously believed worthy of worship, to in fact be repulsive: ugly, disproportionate, and foul-smelling. The value of light and sight is thus made evident.[49]

The association of the Incas with light and sight is brought to the fore in the myth of the battle with the Chancas. Here, Viracocha Inca is presented as representing the traditional, ear-minded order; and Pachacuti, the new, eye-minded one. Whereas Viracocha Inca only hears the voice of Viracocha in the myth, Pachacuti has a resplendent vision of the Sun. The figure of the Sun Pachacuti sees in Molina of Cuzco's version of the myth leaves him a crystal mirror in which he is able to see everything he chooses.[50] This is in striking contrast to the traditional ability of the Andean diviner to *speak* with everything he chooses.[51]

Pachacuti's emphasis on vision and on visual representations is presented in the myth as something new for his subjects. When the leading citizens of Cuzco come to talk with him and see him drawing pictures of bridges and roads, they find this reduction of the world to a flat, visual representation "foreign to their understanding."[52] Pachacuti's structuring ability allies him with the power of light. When Viracocha Inca returns to Cuzco and sees how well ordered the city has become, he declares to Pachacuti; "You are truly the son of the Sun."[53]

The Incas, however, did not wish to negate the power of sound and hearing, but rather to control it. After Capac Yupanqui defeated the huaca through the power of his sight, the Incas were said to have invoked the huacas of their empire to speak on behalf "of the one with strong eyes" (*sinchinani yocpa*).[54] Sight now controlled speech. Significantly, the Inca officials in charge of ensuring that Inca laws were followed in all parts of the empire were known as *tucuy ricuy*, "those who see all."[55]

As Pachacuti structured fluidity, so he structured sound. This seems to be a traditional symbolic role for Inca figures named Pachacuti, or World Reverser. In one myth concerning another Pachacuti figure, for instance, a giant dressed in red (symbolizing turbulence) threatens to cause a flood by blowing on his trumpet (probably a conch shell and thus symbolic of the sea). In this myth Pachacuti is able not only to prevent the giant, who is associated with the chaos of the first age, from flooding the Earth, but also to enter into a harmonious relationship with him.[56] Sound and fluidity are thereby brought under control.

The disorder of sound, again, characterizes the liminal fluid period of the battle with the Chancas. Cieza de León makes particular mention of the confusion of noise at this time. He writes that Pachacuti and the leader of the Chancas met before the battle, but that "they didn't gain anything from seeing each other because they got worked up with words . . . and made a great noise, for here men are great show-offs when they fight and their bark is feared more than their bite."[57]

After the battle, however, Pachacuti is able to control sound as he controls fluidity and make use of its integrating and fertilizing power. Whereas before his victory the inhabitants of Cuzco spoke in discord, now they "all say to him in one voice, 'You are the son of the Sun.'"[58] Similarly, the Incas controlled the discord of the many different languages spoken by their millions of subjects by making Quechua the official language of the empire.

Therefore, although the Incas emphasized the sacred role of light and sight, sound and hearing were still a fundamental subject of symbolic attention. The need for sight to be complemented by hearing was so firmly believed by the Incas, in fact, that they communicated ritually even with the Sun, the preeminent symbol of sight, through sound. Garcilaso de la Vega, for example, writes that after the victory over the Chancas, a messenger was sent to the temple of the Sun "to tell [the Sun] of the victory that had been obtained through his favor and aid, as though he wouldn't have seen it."[59]

5
RITUALS AND INSTITUTIONS

CONQUEST, EXCHANGE, AND THE BODY

The major rituals of the Incas dealt with war, agriculture, and life-cycle passages. Each of these is treated here in turn, in relation to the subject of the human body. The expansionist drive of the Inca state meant that warfare was a standard part of the lives of the Incas and their subjects. All ordinary adult men were subject to military service, and, prowess on the battlefield being richly rewarded, military service was one of the few means by which an ordinary Inca subject could improve his lot in life.

War rituals were primarily concerned with fortifying the powers of the Inca's forces and diminishing those of the enemy. In one preparatory ritual, for instance, a llama was kept without food in the hope that the hearts of the Inca's enemies would become as weak as that of the llama.[1] In an important postcombat ritual the Inca would walk over the prone bodies of his defeated foes proclaiming, "I step on my enemies." So crucial was this rite of stepping on the enemy to the structure of power in Tahuantinsuyu that after the battle with the Chancas, Inca Viracocha and his son Pachacuti had a prolonged dispute over who would step on the Chanca prisoners. Inca Viracocha wanted his chosen heir, Inca Urco, to perform the ritual, but Pachacuti declared that he had not conquered the Chancas "to have them stepped on by women like Inca Urco and his brothers."[2]

It was essential that the person who performed this rite strongly exemplify maleness because the Incas associated their forces with dominant masculinity, and those of the conquered enemy with subordinate femininity. A similar rite to that of stepping on the enemy occurred when Inca Viracocha lay down at Pachacuti's feet and acknowledged Pachacuti as the new Inca. At that time Pachacuti explicitly called his father a woman for failing to defend Cuzco.[3]

War, with few exceptons, was an exclusively male enterprise and an expression of male power. The Inca's troops could not touch women's clothes or possessions in times of war, and women could not touch instruments of battle, no doubt to prevent any debilitation of the army's masculine power through feminine contact.[4]

To lose in war was to be symbolically feminized. This is possibly why the Incas made their prisoners wear long dresses—women's garb—when they were paraded through Cuzco as part of the conquest ceremonies.[5] The ritual of stepping on the enemy asserted the dominance of the Inca social body over that of opposing groups and associated the prone bodies of the enemy with the female qualities of submission, lowness, and earth.

Agriculture, like war, was idealized by the Incas as a specifically male task. In Inca agriculture it was the Earth that was portrayed as the conquered and submissive feminine "other." The way the Incas civilized so-called savage, female outsiders by incorporating them into the Inca order was similar to the way they civilized the so-called savage, female Earth through agriculture.

The agricultural ritual analogous to the rite of stepping on the enemy was the *hailli*, a ritual breaking of the earth with a foot plow, accompanied by song (fig. 6). Garcilaso de la Vega brings out the intrinsic connection between this seemingly simple agricultural rite and Inca rule. "The songs [the Incas] sang in praise of the Sun and their kings were all based on the meaning of the word *hailli*, which . . . means triumph over the Earth, which they plowed and disembowelled so that it should give fruit."[6]

Sex and marriage were also understood to a certain extent by the Incas as the conquest and civilization of "savage" female bodies. The word *purum* could mean an unconquered enemy, uncultivated land, or a virgin woman. Sex and marriage, however, while on the one hand considered fundamental to civilization, on the other hand were seen as a corruption of an original pure state. Thus virginity in both women and men, although not generally prized by the Andeans,[7] could serve to symbolize purity and wholeness within a sacred context.

The Incas created an institution of lifetime virgins called acllas (chosen women).[8] These women were selected at an early age for their beauty from throughout the empire. After a period of training, some were taken or distributed as wives by the Inca, and the rest were cloistered in special houses next to temples of the Sun in order to serve the Inca deities. These acllas were told that if they lost their virginity their bodies would rot and, in fact, were punished by death if they had sexual relations.[9]

Irene Silverblatt writes that this control by the Incas of female

Figure 6. The Hailli Ritual
(From Felipe Guaman Poma de Ayala, *El primer nueva corónica y buen gobierno*, vol. 3, ed. J. Murra and R. Adorn [México: Siglo Veintiuno, 1980], p. 1050.)

sexuality "served to exemplify control over all aspects of social reproduction."[10] The institution of the acllas also had the cosmological purpose of helping to bind the powers of the male Sky to the female Earth by offering women, symbolic of the Earth, to the Sun. As sacred offerings, the acllas had to be whole and flawless; anything else would be an offense to the gods. Virginity therefore could signify sacrality rather than savagery when it supported the cosmological and imperial order of the Incas.

The conquest of the "other," whether people, land, or women, was paired with the establishment of a system of exchange with the other. The Incas, for instance, while imposing their own deities and rites on their subjects, brought the most important huacas of the peoples they conquered to Cuzco, where they were housed in a temple and served by attendants from their communities of origin.[11] Many agricultural rituals consisted of offerings to Pachamama, the Earth, in return for allowing herself to be conquered through agriculture. Wedding rituals expressed the reciprocity involved in marriage. In one ritual, related by Cobo, the groom would give the bride a sandal and receive from her a headband, shirt, and chest plate.[12]

The most fundamental Andean ritual of exchange, and a standard formality of all social occasions, was that of offering a cup of chicha to someone and drinking another cup oneself. This not only was a gesture of hospitality but also served to integrate separate bodies through a sharing of fluids. On the simplest level this ritual would take place between individuals during a social visit. On a more complex level it would seal an alliance between two social bodies, as when an Inca drank with the leader of a conquered people. On a sacred level this rite was practiced between human and sacred bodies: The Sun, the Earth, rivers, and all important huacas were ritually offered chicha in this manner.[13]

Ritual expressions of exchange, as well as affirming the mutual obligations of the participants, contained explicit references to the power structure that underlay them. For example, one would offer the cup of chicha with one's left hand to an inferior, and with one's right hand to an equal or superior.[14] The foreign huacas housed in Cuzco, in turn, were kept chained by the foot in order to ensure their subjugation, and that of their communities of origin, to the Inca.[15] Cobo writes of these huacas that "when a province rebelled [the Incas] ordered its native gods and protectors to be taken out in public and ignominiously whipped every day until the said province was reduced to submission. Once it was so reduced [the huacas'] former places were restored to them and they were honored with sacrifices. . . . It is said that most of the rebels surrendered simply on hearing that their idols were being publicly humiliated."[16]

While conquest was paired with institutions of exchange and integration, therefore, these institutions continued to express the structures of domination the Incas desired to establish through conquest.

RITES OF PASSAGE

Marriage exemplified the Andean cosmological ideal of karihuarmi (man-woman), the union of complementary forces. The basic Inca marriage ceremony was quite simple: The couple would have their hands united by the Inca or else by his local representative. This signified that from then on the couple should function as one, as karihuarmi. This marriage ceremony, which was followed by local rituals, took place once a year, at which time all marriageable men and women were paired off, with more or less degree of consent on their part.[17] It was important that the husband and wife be of equal status and condition, so much so that the blind would be married to the blind, the lame to the lame, and so on.[18] This ensured that the marital relationship would be balanced, and social boundaries not be transgressed.

Marriage was an obligation for all adults, excepting those who were physically unfit or who were sworn to celibacy for religious reasons, and necessary for full participation in adult life.[19] The sexual division of labor, in fact, made life a hardship for any adult without a spouse.

Monogamy was the usual rule for the common people in the Inca empire. The Inca, however, had as many secondary wives as he pleased, or as were useful to cement alliances, and could reward his subjects by giving them extra wives. Nonetheless, even when there were secondary wives, there was always only one principal wife. Women never had more than one husband, as this would have upset the balance of male-female power.

While premarital sex was generally tolerated and trial marriages common in some areas, adultery was a serious crime under Inca law, for it threatened the cohesion of the marital unit.[20] Masturbation and homosexuality were also crimes/sins under Inca law,[21] largely, no doubt, because they "wasted seed," but also because they subverted the ideal of karihuarmi. Homosexuality was particularly abhorred by the Incas and punished with death, possibly because, by presenting the union of like forces as a model, it threatened the order and integration of the dualistically structured cosmos.

Incest was permitted only to the Inca, who was supposed to take one of his full sisters as his Coya (queen). This was in order to maintain the purity of the sacred bloodline and to most closely approximate the ideal of karihuarmi. The Inca and the Coya together embodied the cosmos—the Inca being identified with the Sun, and the Coya with the Moon and the Earth.

The coronation of the Inca, which basically consisted of his receiving the royal insignia and swearing to fulfill his imperial duties, coincided with his wedding. For the Inca, the full adult duties that came with marriage consisted of assuming the leadership of the empire. Conversely, the Inca could not represent the empire without exemplifying the ideal of karihuarmi.

The birth of a son and potential heir to the royal couple was another occasion for public festivities. Such a birth seems to have constituted a liminal period for the Incas during which anomalies were tolerated and even expected. Pachacuti Yamqui reports, for instance, that in the celebration following the birth of one of Inca Pachacuti's sons, luminescent stones and a spotted animal associated with hermaphrodites were brought into Cuzco. As well, the Inca gathered all the dwarves and hunchbacks together to make clothes for his son. On the birth of Huayna Capac, all the "men and women who worship two gods," or sorcerers, who had been sentenced to death were pardoned.[22] This emphasis on anomalies was probably because the birth of a potential heir (among the Incas not the eldest alone, but any son begotten with the Coya was a potential heir) created the anomaly of the coexistence of the present and future Incas.

Few birth rituals are mentioned in the chronicles for ordinary subjects of the Inca. Women would pray to the Moon for a safe delivery, and men were expected to fast while their wives were giving birth. On the fourth day after the birth, relatives would be invited to eat and drink and see the baby in a ceremony called *ayascay*. When the birth was an unusual one, such as of twins, and therefore regarded as a deviation from the normal order, the parents were required to take the ritual precaution of fasting and abstaining from sexual relations for a prescribed period.[23]

When the child was weaned the *rutuchico*, or hair-cutting ceremony, was performed. At this ceremony the child's relatives would each cut a lock of hair from the child's head and in return offer the child a gift. In the case of a child of the Inca these gifts would be very fine; an ordinary child received corn, wool, and like goods. The cutting of the child's hair in exchange for gifts signified the break the child made from its mother on being weaned and its integration into the community. At the rutuchico ceremony children were also given names, another sign of their new social status, which they retained until puberty.[24]

The most important Inca rite of passage (aside from death) was the male puberty rite, called huarochico, or breechcloth ceremony, which took place when boys were about fourteen. In preparation for this ceremony, which was held every year and lasted about a month, Inca boys would fast, have their hair cut, and spend a night on Huanacauri hill beside the huaca of Ayar Ucho, reliving the sojourn of the first Incas. During the ceremonies the boys would participate in rigorous exercises and listen to lectures by their elders. As the culmination of the rite the boys would have their ears pierced—Inca men wore large gold ear ornaments—and receive breechcloths.[25] The length of time and the number of rituals devoted to the male puberty rite indicate the importance the Incas gave to inculcating the new generation of leaders with Inca values.

The puberty rite for girls, *quicuchico*, was of much less importance. A girl would fast for three days in her house from the day she started to menstruate. On the fourth day she would have her hair cut, be dressed in new clothes, and participate in a feast with her relatives, who would lecture her on her adult duties.[26]

During these puberty rites both boys and girls received the names they would keep for the rest of their lives (except in the case of the heir to the throne, who would often take a new name on becoming the Inca). Diego de Castro Titu Cusi Yupanqui says in his sixteenth-century *Ynstrución* that the ear-piercing ceremony is "the most important ceremony we make in the whole year because then we are given names, new names from what we had before."[27]

The puberty rites had the purpose of marking the passage to adult status in the community. The new name given to the youth thus signified his or her new physical and social identity, just as the giving of the first name when the child was weaned signified the child's separate identity from its mother. This new adult identity was confirmed by the change to adult dress that followed the puberty rites. Full adulthood, however, came only with marriage.

AGE GRADES

Aside from these life-cycle rites of passage, the Incas classed their subjects by age, assigning different tasks to different age-groups.[28] There were twelve standard age grades from infancy to old age, with separate classes for men and women. According to Guaman Poma, girls of about nine years, for example, were called *paupau pallac* (flower gatherers) and expected to collect flowers for making dyes, and herbs. Boys of the same age were called *tocllacoc uamracuna* (boy hunters) and had as their task the hunting of small birds.[29] The sexual division of labor was not simply inaugurated by puberty in Inca culture, therefore, but a constant dynamic throughout life.

The most important age grade was that of adulthood, which encompassed the period from about the age of twenty-five to the age of fifty. Guaman Poma lists the duties of men of this age as being to fight in battle, to serve in the mines or other state works, and to populate and cultivate conquered territories. Adult women, in contrast, are said to have as their primary task the weaving of cloth.[30]

While the adult age grade formed the most important labor pool for the Incas, even very young children were expected to help their parents, and the very old to do whatever chores still lay within their capabilities and to serve as advisors. Throughout life, therefore, changes in one's bodily state were directly correlated to changes in one's social state and responsibilities.

The Inca division of human life into age grades may well have corresponded to a degree to the Inca division of the cosmic life into ages.[31] According to such a correspondence, the abrupt changes in one's physical and social identity that occurred with certain rites of aging would constitute a minor pachacuti (world reversal) on a personal level. Having passed through such a rite, an individual, with his or her shorn head, new clothes, and new name, would be metaphorically reborn, and indeed resemble the first humans, who were given names and clothes by Viracocha. In this way the transformation of the human body through age provided a dynamic model of and for the transformation of the cosmos through time.

SACRIFICIAL RITES

Common to all rites of passage, and all important rituals, were sacrifices and offerings to the deities and huacas. The Incas usually used domestic animals in their sacrifices; however, on occasions of great importance humans would be sacrificed as well. This sacrifice was called *capacocha* (royal sin or royal transaction). The victims to be sacrificed were exacted as tribute by the Incas from the four quarters of Tahuantinsuyu. They were customarily strangled, drowned, or buried alive, as it was important that their bodies remain intact in order to embody the Inca ideal of wholeness/holiness.[32] The following description of this ceremony was recorded in 1621.

> It was customary in gentile times to celebrate the feast of the capacocha every four years, choosing four youths from ten to twelve years old, perfect in beauty, without blemish or wrinkle. . . . These four acllas, which means chosen ones, were taken to Cuzco from the four parts of Peru. . . . They entered the plaza where the Inca was seated on his seat of gold together with the statues of the Sun, Lightning and Thunder, and the embalmed Incas and the priests who displayed them. . . . When the feast was over, the capacochas who were meant for Cuzco were taken to the huaca of Huanacauri or to the house of the Sun, put to sleep, lowered into a vault . . . and walled in alive. The others were sent back to their lands to have the same done to them there by order of the Inca, who rewarded their fathers by making them governors, and decreed that there be priests to attend to [the capacocha] and to the rites that were celebrated in its honor every year.[33]

Through the sacrifice of these capacochas, drawn from and symbolic of the four quarters of the empire, the Inca and the empire atoned by proxy for whatever sins they might have accumulated over the past four years. In a similar fashion an individual would commonly offer a small part of his or her body, such as an eyelash, to a huaca as a symbolic substitute for the whole.[34]

The blood of selected animal sacrifices was used to paint a line from ear to ear on huacas and participants in major Inca rites of passage. This ritual, which was significant enough to merit a name of its own, *pirac*, enabled the participants to physically share in the sacrifice and to manifest blood, the preeminent symbol of passage. The horizontal line across the face's natural vertical axis served to order the face into four quarters and thus replicate the quadripartition of Cuzco and the empire.[35]

"Thérèse Bouysse-Cassagne writes of a similar ritual carried out on children among the modern Aymara. "[The horizontal line] unites the two symmetrical halves [of the face] across the axis of the nose: their convergence (*taypi*) establishes a new facial geometry; if at first there were only two halves (right and left), we now have four *tawantin* sectors (right above, left above, right below, left below), a symbolic quadripartition that can be related to the layout of the village. A mutual confirmation is achieved as it were, between the body of the child and the body politic."[36]

The new high/low division created by this horizontal line can be further interpreted as involving the superimposition of a duality based on cultural values on the duality found in nature. The participants in the *pirac* ritual, therefore, are both consecrated by being identified with a sacred offering and incorporated into the social order.

PRIESTS AND HUACAS

Inca rituals were officiated at by priests. The Incas maintained a large and elaborately structured priesthood, with different kinds of priests specializing in different functions; for example, in divination or in hearing confessions. The high priest, called Villac Umu (The Priest Who Tells), held a great deal of power and was always a close male relative of the Inca. While the Villac Umu was apparently elected to office by the Inca nobility, other priests might inherit their positions or else be considered destined for religious life by virtue of manifesting some physical abnormality or having been struck by lightning. Inca priests were usually men, but it was possible for women to perform similar functions at the local level, or within those cults dedicated to female deities, such as the Moon.[37]

Aside from those priests supported by the Incas, there were many local diviners and ministers of huacas throughout the empire. These were tolerated by the Incas as long as they accepted the Inca superstructure and did not use their knowledge to harm others (many diviners, for instance, were experts in the use of poisons).

Priests were commonly required by their office to engage in long periods of fasting and abstention from sexual relations. There were some, however, who particularly dedicated themselves to practices of austerity (fig. 7). Martín de Murúa writes in his chronicle that

> the Incas had some physicians, or philosopher-diviners, who were called *guacacue*. These went naked in the most isolated and gloomy regions . . . and walking alone in these uninhabited places, they gave themselves to divination and philosophy without rest. From sunrise to sunset they constantly stared at the sun, no matter how bright it was, without moving their eyes, and they said that in that bright shining wheel they saw great secrets. They spent all day on foot on burning sands without feeling pain, and they also patiently bore the cold and snow. They lived a very pure and simple life, procuring no luxuries, de-

Figure 7. A Barefoot Inca "Astrologer–Poet" Holding a Quipu in One Hand (From Felipe Guaman Poma de Ayala, *El primer nueva corónica y buen gobierno*, vol. 2, ed J. Murra and R. Adorno [México: Siglo Veintiuno, 1980], p. 829.)

> siring nothing but what reason and nature demanded; [surviving] only on what the Earth produced without being mistreated by tools.[38]

Such persons constituted a third class within the savage/civilized Inca social dichotomy. They had characteristics in common with savages—they did not wear clothes, they did not live in towns, and they did not practice agriculture. However, their purpose in this was not to participate in the chaos and sensuality of the Earth, as the Incas held savages did, but to detach themselves from it. One of the purported abilities of such diviners, in fact, was to sever the yoke of the body to the land, fly through the air, and witness what was happening in faraway places.[39]

Another chronicler speaks of a similar group of diviners who "slept on the ground, ate roots, drank cold water, disciplined themselves with knotted cords," and spent their time contemplating the sun, moon, and stars.[40] He writes that when these diviners passed through a town "all the people followed behind them, for they were held to be holy, and [the diviners] prayed in public for the Inca and his people."[41]

Such diviners, and to a certain extent all priests, were, like the cloistered acllas, outside the normal social order; the diviners did not conquer in the traditional male way, and the acllas were not conquered in the traditional female way. By serving as conduits of the sacred, however, and helping to maintain the Inca at the center of an integrated cosmos, they both served to uphold the social order.[42]

Huacas, requiring the services of a vast number of priests, also manifested a split nature, existing on both the material and sacred levels at once. Huacas were conceptualized as bodies—and, in fact, some actually were mummified bodies—that had to be ritually fed in order for them to use their powers in support of, rather than against, the state.[43]

The larger and more complex the empire became, the greater was the danger of its falling prey to the forces of chaos, and therefore the greater the need to associate the state with the forces of sacred structure (and also of masculinity). By the time of the Spanish Conquest the number of huacas requiring ritual care in and around Cuzco was immense. About four hundred of them—stones, springs, tombs, and so on—were located along imaginary lines, called *ceques*, radiating out from the temple of the Sun. The care of the huacas along these ceques was assigned to different Inca panacas (kin groups), and the ceques themselves probably served as a system to regulate social relations among the panacas, thus allying sacred structure with social structure.[44]

Some of the huacas along the ceques were related to water, and the ceques might also have served to ritually direct the flow of water in and out of Cuzco (as in the *citua* rite discussed in the next chapter). In fact, the ceque system was complemented by an underground system of canals that carried chicha offered to the huacas to the different temples of Cuzco.[45] This dual function of ceques, to separate into categories and to channel fluidity, perhaps explains why the word *ceque,* according to González Holguín, means both "borderline" and "watery chicha."

Here then, we find the creation of land boundaries and irrigation canals, attributed by tradition to Inca Pachacuti, replicated on a ritual plane. Sacred bodies, the body politic, and the body of the land are seen to be structured and integrated by similar and interrelated systems that, in turn, correspond to the model of structure and integration provided by the cosmic body.

SIGHT AND HEARING

The ceque system of Cuzco was also a system of sight lines. Zuidema describes how this system worked on one ceque along which he was able to locate many huacas by using information found in the chronicles. "The seventh huaca [of the ceque] on the horizon from Cuzco was used for astronomical observation. The first six huacas lie under the sight line extending from the temple of the sun to this point. By way of the next three huacas, the ceque then passes over the nearby hills. From the ninth huaca on, there is again a sight line to the last (thirteenth) huaca in the new valley."[46]

The use of these huacas as visual markers is also brought out by Cobo, who describes certain huacas along the ceques as being spots where one first catches or loses sight of a sacred place. For example, the ninth huaca on one ceque "was the place where those who were going to Chinchasuyu lost sight of Cuzco."[47]

According to Zuidema, the ceques always ended just beyond the immediate horizon of Cuzco.[48] He finds the horizon also stressed in the Inca ritual of painting a line of blood across the face, during which, according to one chronicler, a line of blood was also painted on the temple of the Sun.[49] Zuidema comments that, "actually, we know that such a line existed at the height of the windows on the inside of the rooms of the temple in Cuzco, and that it was made not of blood, but of gold. The windows were used for observing sunrise or sunset at specific days of the year. I suggest, therefore, that painting blood from ear to ear was intended to stress the eyes; making the eyes dominate, through their sight, the area towards the horizon where the sun rises and sets."[50]

The ceques, as sight lines, may have served to sanctify and structure that area of Cuzco between sunrise and sunset, protecting it from the forces of darkness beyond. Astronomical sight lines, some of which coincided with ceques, in turn made it possible not only to structure time and coordinate rituals with the movements of the heavenly bodies, but also to direct sacred power from the sky to the Earth along the line of vision.

For the Incas, looking at the sky, and thereby transmitting personal power upward, invited a corresponding transmission of power downward. This occurs, for example, in the cosmogony when Viracocha causes fire to fall from above by looking up at the sky. Another example comes from after the Conquest, when a group of Inca rebels buried a Spanish priest, whom they had attacked, upside down so he could not look at the sky. Murúa explains in his chronicle that "the barbarians thought that if they buried him head up God would hear him and take him out of the hole, and so they buried him head down so that he would be unable to lift his eyes up to the sky and call God."[51] For Murúa, such logic was no logic at all. He proclaims the Indians "blind men, without reason or understanding."[52] For the Incas, with their emphasis on reciprocity, it made perfect sense. In the words of Pachacuti Yamqui's prayer to the Creator, "by seeing you . . . you will see me."

Looking at the sacred, however, was considered by the Incas to be a potentially dangerous experience. An ordinary subject would not dare to look at a capacocha sacrificial procession, for example, or at the face of the Inca.[53] Even priests when ministering to huacas would customarily keep their eyes downcast.[54] This vulnerability of sight made the Incas' reputed domination of the huacas by sight truly a feat of exceptional power.

The ceque system can also be compared to a quipu, the Inca device for recording data.[55] The quipu consisted of a main cord from which threads with knots representing data were hung (see the example in fig. 7). In the ceque system the ceques were like the spread-out threads of the quipu, and the huacas like the knots. The orality of the huacas was ritually controlled by the sight lines of the ceques, just as oral knowledge was visually structured by the quipu. Locating the huacas along sight lines, indeed, was probably an important means by which the Incas dominated these oracles through sight.

In Inca ritual, as in Inca myth, fluidity and sound were often related. When the ritual involved extraordinary fluidity, either actual or metaphorical, an extraordinary outburst of sound would accompany it. In the Inca rain ritual, for example, rain would be invoked by a public outcry in which even dogs and llamas were made to participate.[56]

The piercing of ears that took place during the male puberty rite was also associated with sound and fluidity.[57] This rite was the male counterpart to first menstruation and thus an inauguration of fertility. However, the fact that the organ made to bleed was not the penis but the ear suggests that men were supposed to control their sexuality by listening to and obeying oral tradition. Indeed, at the close of the ceremony the boys received ear ornaments, which emphasized the ears and hearing, and breechcloths, which covered the genitals. Piercing the ears, then, can be understood as both inaugurating and controlling male fertility, and opening up the ear to be fertilized by sound, in particular the oral communications of the sacred.

Significantly, while the ears might ordinarily attract less attention than the other organs of the face, the Incas highlighted them by keeping their hair short and wearing distinctive large gold ear ornaments. This drew attention to the Incas' privileged access to sacred orality.[58] The Incas were the ones with the ability to make the huacas speak, and to hear what they said[59] (fig. 8). One Inca myth, recorded by Cieza de León and abbreviated below, draws attention to this special ability of the Incas and intimately relates the ear-piercing rite with sound and fluidity.

> When Inca Roca (the sixth Inca) had his ears pierced, one of them hurt so much that he went to a hill near Cuzco seeking relief. While on the hill he prayed to the Inca deities for help in providing Cuzco with a river, for at that time the city had none. As he was praying he heard a clap of thunder so loud that he lay down with fright, putting his left ear, from which blood was gushing, to the ground. Suddenly, the Inca heard the noise of running water under the earth. He ordered the spot dug up until a spring was found and then had a canal built to carry the water to Cuzco.[60]

The body of the Inca seems here to be a metaphor for the land. His earache is equivalent to the lack of water suffered by Cuzco. There-

Figure 8. Tupac Inca Talking with the Huacas
(From Felipe Guaman Poma de Ayala, *El primer nueva corónica y buen gobierno*, vol. 1, ed. J. Murra and R. Adorno [México: Siglo Veintiuno, 1980], p. 235.)

fore, when Inca Roca goes to the hill he prays not for relief for his ear, but for a river to relieve Cuzco from its drought. The blood that then flows from his ear corresponds to the river he hears flowing underground.

On another level, the Inca's earache can be understood as resulting from a lack of sacred oral communication, symbolic fluidity. This metaphorical drought is relieved by the thunderclap that comes in response to Inca Roca's prayer. With his newly pierced ears, Inca Roca hears both the sacred thunderclap—signaling the presence of water—and the water running underground. The conclusion is that Cuzco is provided with an irrigation canal, and the Inca establishes a channel of oral communication between himself and the deities.[61]

Inca rituals were usually rich in both auditory and visual phenomena. Music and recitations were essential parts of most rituals, as was a colorful display of costume and ornamentation. The Incas had an elaborate color vocabulary (see the appendix), and undoubtedly the colors used in their ornamentation and rituals had symbolic value. Pachacuti Yamqui, for instance, says that the highest order of acllas were called *yuracaclla* (white acllas) while a middle-ranking order was called *pacoaclla* (red acllas), and the lowest order, *yanaaclla* (black acllas).[62] Murúa writes that spotted llamas were sacrificed to Thunder, brown llamas to Viracocha, and white llamas to the Sun.[63] However, while the colors used in different rites by the Incas were often recorded by the chroniclers, their meanings, unfortunately were not.

The uniting of the auditory and the visual was fundamental to Inca ritual. One example of such a union was the use of *runa tinya*, human drums made out of the bodies of traitors. The whole corpse would be preserved in its native costume, but the belly would be made into a drum that would be played with the corpse's own hand.[64] It was evidently not enough for the traitor's body simply to be displayed, it had to be heard as well. Whereas before the traitor spoke against the Inca, now he was made to "speak" in support of the Inca.

The ritual union of sight and hearing is best exemplified by a harvest thanksgiving ceremony in which the Incas would go before dawn to a plain to the east of Cuzco. As the sun began to rise they would start to sing in chorus, with the Inca leading. The higher the sun rose, the louder they sang. When the sun set they sang more and more softly, stopping when it disappeared from sight.[65]

To have merely observed the sun rising and setting would not have been ritually complete for the Incas, they had to add an auditory dimension to light in order to assimilate it into the oral structures of their culture. In consideration of this, perhaps the line of blood painted across the face in certain rites of passage also had the symbolic purpose of uniting the eyes with the ears and, consequently, the seen with the heard.

TASTE, SMELL, AND TOUCH

The emphasis the Incas placed on sight and hearing in their cosmology and rituals did not preclude the involvement of the other senses. Even the most visualist of Inca practices, observing the stars, was intimately related to the different sensory perceptions produced by the characteristic activities, songs, foods, and so on, corresponding to the different seasons. In his book on Andean astronomy Urton writes that "it is important to note that the divisions and passage of time are related to different crops and activities, and that these in turn are related to the different senses. Since the astronomical reckoning of time depends primarily on vision, [this] suggests that the total perception of time and space will involve the union of all sensual perceptions of change in the environment."[66]

Inca rituals, indeed, commonly engaged all of the senses through colorful pageantry, songs and dances, feasting, and the burning of incense. Most, if not all, of these sensory phenomena probably had symbolic meaning. Garcilaso de la Vega, for instance, writes of the wreaths of flowers placed on the heads of the Inca youths at the conclusion of the huarochico ceremony that "the beautiful and fragrant flowers were said to signify [the Inca's] clemency, compassion and mildness."[67]

Food and drink were integral to many Inca rituals, not only as part of the postritual celebrations, but within the rituals themselves. The most obvious example of this is the ritual drinking of chicha in order to confirm a relationship of reciprocity. Another example is the ritual eating of *yahuar sanco*, balls of cornmeal mixed with the blood of sacrificed llamas, following the swearing of an oath of allegiance to the Inca.[68] In such ceremonial use of food, the emphasis was on the incorporation of the ritual substance and the concepts associated with it into the body.

For the Andeans, to eat something was, to a certain extent, to become one with it. For this reason, for instance, a person who wanted to cross a river safely would drink some of the river's water.[69] Speaking certain words was thought to have a similar effect. Garcilaso writes that women would never say the word for making sandals because to do so would be to make themselves into men, as making sandals was men's work. For similar reasons, men would never say the word for weaving, the principal occupation of women.[70] In both the oral functions of eating and speaking, therefore, the sign (food or words) was identified with the thing signified, and considered capable of producing a transformation in the person who ingested or uttered it.

Flavors, however, also had a metaphorical significance. The Quechua words for sweet, *misqui* and *ñucñu*, had connotations of mild, good, and pleasant; while the word for bitter, *hayak*, also meant "harsh and unpleasant," and surprisingly, in certain contexts, "excellent." The expression *hayaqquentarimani* (to speak bitterly), for instance, meant "to speak admirably." González Holguín comments that, "you would think it would mean the opposite because *hayaqquen* means 'gall or bitterness,' but that is how they speak."[71] This might be because the Andeans considered bitterness to produce an awareness of categories and distinctions necessary for a true understanding of reality.

The other two Quechua words that could be employed in a similar way as hayak were *huañuy* (deadly) and *millay* (disgusting). González Holguín writes that these were used as adjectives to mean "entirely" and "perfectly," as, for example, in the expression *huañuy yachak* (deadly wise). This suggests that a bitter taste, death, and a disgusting sensation were all thought to produce such a sharp break in the continuity of existence as to have a supra-real quality of absoluteness. How and if such linguistic expressions were translated into ritual by the Incas, however, is impossible to determine.[72]

The chronicles have little more to say about the role of odor in Inca ritual than they have about that of flavor. The large number of words in Quechua relating to odor, however, indicates that the Incas were not inattentive to the sense of smell (see the appendix). Inca sacrifices were always burnt on scented wood in order to make them more pleasing to the deities, and huacas were customarily perfumed with incense.[73] Conversely, foul odors were used by the Incas in order to repel. Thus the priest who was buried upside down by the Inca rebels was made to smell a foul incense that "took his breath away and left him speechless."[74]

The most repugnant odor for the Incas was that of putridity. Indeed, to call someone rotten-smelling was one of the worst insults an Andean could proffer. The inhabitants of Collasuyu, for example, were called stinking (*aznay*) by the Incas because they were very rebellious.[75] This probably had to do with the association of putrefaction with disease, death, and dissolution, and, by extension, with anticulture and savagery.

According to Murúa, the acllas were supposed to live on odors alone. "[The acllas] have no need of food, they live only on the odor of a certain wild fruit, and when they are traveling away from their house they take some of this fruit with them in order to sustain themselves with its odor. If they were ever to smell a bad odor, it was certain that they would die without remedy."[76] This myth was undoubtedly intended to emphasize the spirituality and exquisite sensitivity of the acllas. Significantly, the odor the acllas are said to live on comes from a *wild* fruit, associating the acllas with the unconquered Earth and putting them again on a similar plane to the diviners who lived off the uncultivated products of the Earth.[77]

The acllas, like the diviners, maintained their spirituality in great part by remaining chaste or, in other words, restricting their sense of touch. Of all the senses, touch was the one the Incas considered most able to break down the barriers between categories, and therefore the one that had to be most rigorously controlled. Everything the Inca touched, in particular his garments, for instance, was kept in chests and burned every year so that no one else could touch it and thereby transgress the barriers that separated the Inca from his subjects.[78]

The Incas were accustomed to have their sense of touch blunted from birth. According to Garcilaso de la Vega, babies were washed in cold water, exposed to the night air, and kept swaddled in cradles until they were three months old (fig. 9). Mothers never held their children in their arms or laps, not even while breast-feeding, because they believed such pampering would make them crybabies.[79] Boys in training for the huarochico rite were similarly hardened by being made to sleep on the ground, suffer pain without flinching, and take part in other exercises designed to prepare them for the rigors of military life.[80]

This cultivated insensitivity to pain was contrasted by the very soft clothing worn by the Incas. As only the Incas were allowed to wear fine, soft clothing, while that of their subjects was coarse and rough, this distinction in textures served as a tactile expression of Inca class distinctions.[81]

Figure 9. A Swaddled Inca Baby
(From Felipe Guaman Poma de Ayala, *El primer nueva corónica y buen gobierno,* vol. 1, ed. J. Murra and R. Adorno [México: Siglo Veintiuno, 1980], p. 188.)

Touch was engaged by Inca rituals in various ways. In the huarochico rite the Inca elders would whip the initiates as they lectured them on their responsibilities,[82] possibly in the belief that the elders' words would thereby physically penetrate the boys' bodies, just as the oath of allegiance to the Inca was physically incorporated through the eating of yahuar sanco.

Dance, an essential part of all Inca ceremonies, engaged both touch and the kinesthetic sense, along with sight and hearing. The Incas had a number of ritual dances that were danced in groups following a rhythmic pattern. Molina of Cuzco describes an Inca dance called *yavayra* in which the men would hold one end of a long rope made of black, white, red, and yellow wool, and the women the other. Following the leader, the dancers would thread through the streets of Cuzco paying homage to the huacas and the Inca. At the end of the dance the rope would be left coiled on the ground of the plaza like a snake.[83]

This dance followed a ritual battle between Hanan and Hurin Cuzco and can therefore be interpreted as a ritual reintegration of the two moieties of the city. The antagonistic duality of the battle became complementary duality in the dance in which men and women, while at opposite ends, were united by holding the same rope and following the same steps. The four basic colors entwined in the rope, in turn, represented the principle of integrated quadripartition. The rope also stood for a snake, an Inca symbol of divine fluidity, and the weaving of the rope-snake through the streets of Cuzco helped accomplish the fluid integration of the city.

Other Inca dances served similar purposes: to enact cosmological and imperial principles; establish communion with the sacred; effect the passage from one state to another; and affirm the social order. For this last purpose the Incas sometimes ritually incorporated the bodies of slain enemies into a dance. Guaman Poma quotes the Incas as saying, "We will drink from our enemy's skull, we will wear his teeth as necklaces, we will play his bones as flutes, his hide as a drum, and thus we will dance."[84] This transformation of the bodies of enemies into the ritual elements for a dance was an enactment of order over disorder and a dynamic incorporation of rebellious bodies into the Inca social body.

Dance, perhaps because of the continuous movement of the dancers as opposed to the stillness of death, was identified by the Incas with life itself. Guaman Poma writes that in order to counteract omens prognosticating death the Incas would hold night-long dances. "They say that [the omen] is a very bad sign and that they are going to die and be finished, and so as not to die they dance all night."[85] The body here is no longer a static axis through which divine power flows, but divine power itself, animating the cosmos as it courses through it.

SENSORY DEPRIVATION

Inca rites of passage were customarily preceded by a short period of sensory deprivation, usually consisting of abstaining from condiments, tasty foods, and sex. In the case of a person filling a religious office, the deprivation could be quite rigorous. Arriaga tells us that "a person about to assume one of the [priestly] offices has to fast for a month, or in some places a year, eating neither salt nor pepper, nor sleeping with his wife, nor washing nor combing his hair. In some localities they are forbidden to touch the body with their hands."[86]

This asceticism can be seen as a form of ritual exchange: One deprived oneself of certain things in order to receive other desired things—divine powers or a good harvest, for instance. Significantly, the deprivation was principally one of sensory pleasure, with the senses particularly affected being those apparently considered the most productive of sensory pleasure—touch and taste. Fasting for the Andeans, for example, consisted not of abstaining from food but of abstaining from flavor. This deprivation of sensory pleasure would leave the body receptive to the sensory expressions of ritual and of the sacred. It may also have served to protect those persons whose bodily boundaries would be abnormally weak because of their liminal position in between categories from being disintegrated by dangerously strong and intrusive physical sensations.

Those Andeans who engaged in the most extreme practices of asceticism were the diviners who lived alone in the wilderness, slept on the ground, and ate roots. In the case of diviners the blunting of the proximity senses may have been intended to increase the sensitivity of sight and hearing, the primary media of divine revelation. Murúa, for example, mentions that the ascetic philosopher-diviners gained their knowledge from staring at the sun all day long.[87] Other diviners, who were in charge of the huacas inside the temples, went so far as to blind themselves in order to privilege their sense of hearing and attend to the oral communications of the huacas.[88]

SYNESTHESIA

As in Inca myth, therefore in Inca ritual sight and hearing were stressed above the other senses. However, while Inca myths commonly made no mention of taste and smell, and little mention of touch, in Inca ritual these proximity senses were all actively engaged. Perhaps this is because, as mentioned earlier, in Inca ritual it was considered necessary to involve all of the senses in order for the ceremony to be completely effective.

Some scholars have defined this ritual engagement of several senses at once as "synesthesia." Sullivan, for instance, writes that ritual synesthesia is "the symbolic experience of the unity of the senses [that] enables a culture to entertain itself with the idea of the unity of meaning."[89] Similarly, Edmund Leach has stated that since, when we experience the world, "we are aware of a *single* total experience . . . it must be because the coding of the various sensory signal systems can be consistent so that hearing *and* sight *and* smell *and* taste *and* touch, etc., seem all to be giving the same message."[90]

Rather than all conveying the same message, however, the senses in Inca ritual seem to have conveyed complementary messages, according to what the Incas understood the particular characteristics of each sense to be. Thus sight would be used to structure, hearing to integrate or transform, and so on. Furthermore, the sequencing and patterning of sensory perceptions was undoubtedly intrinsic to Inca rituals, carefully ordered and encoded as they were.

The capacocha ritual provides a good example of such sequencing. The sensory stages of this ritual were as follows. In the first stage the sacrificial victims circled the central plaza of Cuzco, where the Inca was seated together with statues of the deities and his embalmed ancestors. This provided a visual affirmation of the structure of the cosmos and empire. In the second stage the Inca spoke to the Sun, asking him to accept the chosen victims. Sound was used here to open up a channel of communication and exchange with the divine. The establishment of this channel was confirmed by the Inca's exchanging a toast of chicha with the Sun, thereby supplementing sound with fluidity.

In the next stage the Inca rubbed his body with either the dregs of the chicha offered to the Sun or with the victims (the meaning of the Spanish text is unclear), in order to effect a transfer of divinity. Touch was thus used to transcend physical barriers and enable either the Inca to participate in the divinity of the Sun, or the victims to participate in the divinity of the Inca.

The last public stage of the ritual consisted of the Inca priest's offering sacred yahuar sanco balls along with pieces of meat from the sacrificed llamas to the participants to eat. This final ingestion of sacrality involved the sense of taste. When the feasting, which lasted several days, was over the victims were taken away and sacrificed.[91]

Thus at each stage of the ritual a different sense was emphasized to a different purpose, with a progression taking place from the exteriority of sight to the interiority of taste.[92] This progression took place within the more general framework of preritual fasting followed by postritual feasting, a transition from sensory deprivation to sensory fulfillment. As, in Inca mythology, salt and pepper (personified as Incas) appear with the birth of the Incas and are associated with Inca order, this preritual fasting followed by postritual feasting was perhaps also a symbolic reenactment of the original mythological transition from nature to culture.

Synesthesia, in its more specific sense, refers to the translation of a perception belonging to one sensory modality into another sensory modality; for example, experiencing a sound as a color. The consumption of the yahuar sanco balls, which embodied an oath of allegiance to the Inca, may be taken as an instance of such a sensory translation in Inca ritual. However, the records we have of Inca ritual (and myth) contain few references to such synesthetic practices.[93]

The implication to be drawn from this is that, while the Incas were concerned to symbolically pattern and combine sensory perceptions in their rituals, they did not want the boundaries between the senses to be completely obliterated. This ordering of the senses would correspond to the Inca ordering of the cosmos in which the categories of male and female, structure and fluidity, and so on, were complementary, but not interchangeable.

6
ILLNESS AND DEATH

ILLNESS

The Incas conceived of illness as a harmful presence within the body. Inca curers, called in general *hampiyok*, used various methods for removing a disease from a patient's body. One of the most effective was herbal infusions, of which the hampiyok had an extensive knowledge. Other methods included sucking the disease out of a patient, bloodletting, and leaving the patient's clothes by the roadside so that the illness would be carried away by whomever came by and picked the clothes up.[1]

According to the Incas, illness was often caused by *hucha*, or sin, an element that disrupted the unity of the whole. As the Incas tended to identify the moral constitution of a person with that person's physical constitution, it was natural that a moral flaw should be thought to produce a corresponding physical defect. The moral flaw itself was considered to be either structural and usually innate, or nonstructural and produced by a breach of custom. Structural hucha, which manifested itself in physical deformities, could not be removed. Nonstructural hucha, which caused illness and misfortune, could be removed through confession, penance, and ritual cleansing.

Inca subjects were expected to confess all their sins to confessors, called *ychuri*. There would usually be one ychuri attached to an ayllu, or community. Confessions were made a few times a year during festivals, or when a person suffered some misfortune, such as illness, considered to have been produced by sin. Juan Polo de Ondegardo writes that "it was held to be very sinful to hide any sin in the confession, and the ychuri, or confessors, check by casting lots or looking at the entrails of an animal whether some sin is being hidden from them. They punish [the person who has hidden a sin] by hitting him on the back with certain stones until he tells everything, and then they give him a penance and make a sacrifice."[2]

One method an ychuri used to determine whether a confession was complete was to divide a handful of straw into two parts and then count the straws on each side to see whether the piles were even. If they were, the confession was held to be good; if not, it meant that there were still sins left unconfessed. The word *ychuri* itself was derived from the name of this straw, *ychu*, and means "one who picks up straw."[3] On confessing, a person would spit into this straw.[4] The sin that was thereby expelled might thus be thought of as a corrupting fluid within the body.

After confession an offering would be made to the deities or ancestors offended by the hucha, and a penance imposed on the guilty party. The most common penances were to abstain from salt, pepper, and sex for a period, or to be whipped. This last was supposed to be done by a hunchback or a person with some other congenital deformity.[5] The reason for this was perhaps to stress the innate monstrosity of sin and its disruptive effects on the body.

As a conclusion to the confession, the hucha would be washed away by a ritual bath in a stream, and preferably the juncture of two streams to better disperse the sins. These baths were called *opacuna*,[6] which means "deaf or dumb ones," referring possibly to the metaphorical speechlessness of those who have had their confessions carried away by the water.

The Inca confessed directly to the Sun. As he washed his sins away in the river he would say, "I have told my sins to the Sun; you, river, take them to the sea where they will nevermore appear."[7] When the Inca was sick all of his subjects confessed,[8] a practice that confirms that an identification was made of the body of the Inca with the body politic. The health of the Inca, representing as it did the health of the empire, was of the highest importance for the Incas and the subject of many rituals. Many of the huacas situated on the ceques, for instance, had as their primary purpose the conservation of the Inca's health.[9]

The health of the Inca social body was directly safeguarded by an annual purificatory rite called citua. This rite was held in Cuzco before the start of the rainy season, a time when Andeans were particularly susceptible to disease. All foreigners and persons with physical defects had to leave the city as a necessary prelude to the purification of Cuzco. For the performance of the rite four groups of Inca warriors would gather in the plaza, each group facing a different direction. At the appearance of the new moon, Inca priests would leave the temple of Coricancha calling out, "Illnesses, disasters, and misfortunes, leave this land!" On hearing their voices, the groups of warriors would start running, each group in the direction it faced, and shouting out, "Evil begone!" These warriors would be replaced by others along the way who would continue running until they reached certain rivers where they washed the evil from their bodies. It was believed that the evil was then taken away by the rivers to the sea.[10]

As part of this ceremony all the inhabitants of Cuzco would bathe, each clan in springs or rivers located along its own ceque. After this they would rub their faces, doorways, and storehouses with ground corn and then throw the corn into springs in a ritual designed to protect them from disease. Even the Inca mummies would be ritually bathed and rubbed with corn in this manner.[11]

In the festivities that followed a special dance took place in which the dancers wore long red gowns and played panpipes (*antara*). When the people who had been sent out from Cuzco before citua were allowed back in after the rite, they were given yahuar sanco to eat "as a sign of their confederation with the Inca."[12] Once this sacred food was in their bodies it was believed that it would reveal any crimes committed by them against the Inca.[13]

Citua was intended to purify the body of Cuzco as a whole and the individual bodies of its inhabitants. The warriors ran in the four directions of the empire in order to cleanse all four quarters of the city and disperse the illness in all directions. As in confession, the evil was expelled both orally, by shouting "evil begone," and by water, through ablutions. The fluidity of both sound and water were thus necessary to effectively cleanse the body.

There may well have been a symbolic connotation of menstruation in the citua ritual. The rite took place in the month of Coya Raymi (Festival of the Queen), which was dedicated to the Moon and to women and therefore associated with feminine cycles. In this regard it is interesting to note among the modern Qollahuayas that, "because of menstruation, it is said, women are better able than men to get rid of bad things. The flow of blood cleanses a women of misfortune, such as infertility, and prepares her for conception."[14] In the dance performed after citua the red-gowned dancers would act out this cleansing flow of blood from the body. At the same time the panpipes played by the dancers added the necessary auditory component to the fluidity expressed by the red gowns and the dance.

The illnesses the Incas sought to be rid of at citua included both physical and social disorders. In fact, the sins—adultery, theft, murder, and so on—that were believed to cause physical illness in the body of the individual who committed them usually caused a corresponding "illness" in the social and cosmic body.[15] The festival of citua, intended as it was to rid Cuzco of social as well as physical illness, was therefore one of exemplary social behavior for the Incas. Accordingly, "they did not reprove each other during that time, nor speak angry words, nor ask for debts to be paid, because they believed that if they were angry or quarrelsome on that day, they would be that way the whole year."[16]

Citua can be regarded as ritually returning Cuzco, and by extension the empire, to its original pristine, ideal state. For this reason all foreigners and persons with physical defects had to leave the city before the rite was performed and had to be ritually reincorporated into the Inca order when they were allowed back in, by eating the yahuar sanco, symbol of Inca holiness. The reason citua was performed before the start of the rains may have been not only that the rainy season was a time of sickness, but also that it was necessary to purify and fortify bodily and social structures lest they be dissolved by the excessive fluidity.

In the rite of citua, as in confession, water was used to carry illness away. Water, however, was also thought to have the ability to carry away strength. The spring in which the Inca bathed, for instance, had to be ritually placated so that it would not rob the Inca of his vitality.[17] Bastien writes about a similar belief among the Qollahuayas. "Juan Wilka, a Kaatan curer, said that Elsa Yanahuaya's blood was weak because a landslide had taken it and replaced it with water. Landslides, floods and turbulent streams wash the land away; and water, instead of blood, flowing through the body is associated with the loss of land—as well as with death."[18] The rite of citua can be understood as an attempt both to rid Cuzco of disease and to control water's power of dissolution by directing excessive fluidity away to the sea.

The loss of vitality caused by bathing in springs, known as *pucyo oncuy* (spring sickness), is one of the illnesses Guaman Poma says the Incas sought to rid themselves of at citua. The other illnesses mentioned by him as being expelled in citua are *taqui oncoc* (dancing sickness), *sara oncuy* (corn sickness), *pacha panta* (horizon sickness), *chirapa uncuy* (rain with sun sickness), *pacha maca* (grasp of the earth), *acapana* (red clouds), and *ayapcha oncoycona* (cadaver sicknesses).[19] What these sicknesses all appear to have in common is that they are caused by an anomaly or liminal situation of one kind or another: water bubbling out of the earth; the horizon in which earth and sky meet; rain with sun; the red clouds of twilight, and so on. Like hucha, these conditions transgress categorical boundaries and thus were thought potentially dangerous to one's bodily integrity. When the Incas drove out illnesses from Cuzco in citua, therefore, they were also ridding their body politic of ambiguities and reaffirming its structural divisions.

DEATH

Although the word *deadly* (huañuy), was used as an adjective by the Incas to mean "entirely and perfectly," death, in fact, did not signify the absolute end of human existence in Inca cosmology. George Urioste remarks that "preconquest Andeans thought of death not as the end of life but as a new stage in the process of becoming. The Quechua word for dead people is *wañuq* which means those who are dying, or those in the process of death. However, dead animals are said to be *wañusqa,* or having been affected by death as a permanent attribute. The active character of the human dead explains why they are considered full participants in the social structure of living Andeans."[20]

The rituals that marked death among the Incas and their subjects had points in common with those of other Inca rites of passage. When the deceased was a man his female relatives, and particularly his wife, would cut their hair to indicate their altered social status. So closely identified was this act with death that Guaman Poma writes that if a woman dreamed her hair was cut, it meant she was going to become a widow.[21] The relatives of the deceased also marked the event by wearing black. As in other rites of passage, the cutting of hair and change of apparel were both a visual manifestation of the altered status of the bereaved and a physical assumption of the new status.

Dance and an outpouring of sound—weeping, singing, music—were essential to the mourning rites. They served both to express grief and to help accomplish the transition from one stage in the life cycle to the next. The reciting of major events from the deceased's life was a particularly important means of accomplishing this transition. "[The mourners] went out every day and danced to drums and flutes and sang in sad tones. They went around all those places that the deceased used to frequent while alive, telling through their songs all the things that had happened to him in his life.[22] This took not only the mourners step by step through the stages of the deceased's life, but also the deceased, who might otherwise not be able to make the passage from life to death. At the same time the association between the deceased and the land he or she had lived on was reinforced. Similar rites exist in the modern Andes; for example, among the Qollahuayas: "As Guillermo [who is dying] named the places which he had plowed, planted, and harvested since he was four years old, everyone relaxed and mentally made the journey with him through the lower and rotative fields. They were happy . . . to know that Guillermo was associated with so many places. By cultivating these places, Guillermo had become a part of the mountain, which would remain forever."[23]

The Incas believed that there existed a reciprocal relationship between the living and the dead, whereby each contributed to the well-being of the other and the stability of the cosmos. The contribution made by the dead was to use their power as huacas to safeguard their descendants. The living, for their part, provided the dead with worship and offerings.

As the dead were presumed to have needs similar to those of the living, the most important offering made to them was food and drink. Arriaga writes that the relatives of a deceased person "pour chicha discreetly into the tomb so that the dead [person] may drink, and they make a show of doing him honor, placing cooked meals and roasts upon the grave for him to eat."[24]

New clothes and utensils were also buried with the dead for their use, and in the case of the Inca favored wives and servants would be ritually killed to accompany their lord in the afterlife.[25] The period of mourning lasted up to a year, depending on the rank of the deceased. Offerings to the dead were made not only during this period, however, but at regular periods thereafter.

Guaman Poma states that among the Incas the feast of the dead took place in the month called Aya Marcay Quillay (Month for Carrying the Dead), roughly equivalent to November (fig. 10): "In this month they take the dead out of their vaults, called *pucullo,* and they give them food and drink and dress them in fine clothes and put feathers on their heads and sing and dance with them. They put them on litters and carry them from house to house through the streets and the plaza. Afterward they put them back in their pucullos, giving them their food, with dishes of silver and gold for the nobles and clay for the poor."[26]

Guaman Poma adds that this was also the month in which life-cycle rites in general were celebrated. This suggests that it was thought to be the most conducive period for effecting transfers

Figure 10. The Feast of the Dead
(From Felipe Guaman Poma de Ayala, *El primer nueva corónica y buen gobierno,* vol. 1, ed. J. Murra and R. Adorno [México: Siglo Veintiuno, 1980], p. 230.)

from one category to anoher. The fluid nature of this month, which fell in the rainy season, was perhaps one reason why, according to Guaman Poma, it was also the time when the Incas exacted tribute from their subjects and distributed goods and wives among them.[27]

The bodies of the dead in the Inca empire were customarily kept in vaults where they were accessible to their descendants. When European burial practices were introduced after the Conquest, Andeans would consistently disinter their dead and reinter them according to traditional practice. Arriaga writes of this that "their greatest abuse is to disinter the dead and remove their bodies to their *machays*, or burial places of their ancestors in the fields. . . . On being asked why they do this, they say that this is *cuyaspa*, for the love they bear them."[28]

Exactly how the Incas conceived of the dead is difficult to determine. Apart from the corpse itself, a deceased person was apparently considered to have a spiritual body that would partake of the essences of food and drink offered it and even leave footprints behind in the ashes or cornmeal scattered by relatives around its former home. This spirit was thought to wander about for a period after death and then retire to its permanent home, whether in the sky or in the Earth.[29]

In a tradition that may well be indicative of pre-Conquest beliefs, the Andeans of present-day Cuzco distinguish three souls in the person: the *kuraq* (great soul); the *chaupi* (middle soul); and the *sull'qa* (little soul). These souls separate at death. The kuraq goes to the mountain peaks, the sull'qa transmigrates to another living body, and the chaupi remains with the corpse.[30]

The Incas located reason and the emotions in the sonco, or heart and stomach, and memory in the head, or uma. Camay (breath or inspiration), might have been considered a kind of animating force providing life and linking a person with the spiritual aspects of the physical environment, while the term *nuna* apparently referred to a person's spirit.[31]

Physical sensation was believed by the Incas to reside in the bones, *tullu,* and to remain there after death. For this reason, for example, the post-Conquest Inca, Manco Inca, when dying, beseeched his son not to "make his bones suffer" by mistreating family members.[32] This belief made the Inca ritual use of the bones of enemies as drinking vessels and flutes more than simply symbolic.

If the living could harm the dead, the dead could also harm the living, causing them misfortune and disease. This would occur when some aspect of ancestor worship was neglected, but also spontaneously through contact with a corpse or its emanations. This last is referred to by Guaman Poma as ayapcha oncoycona (cadaver sicknesses),[33] and is generally known in the modern Andes as *machu wayra*, ancestor wind. It is believed to cause a variety of sicknesses, but principally one that starts off as a rash, then enters the bones and destroys the whole body.[34] Contact with a corpse therefore seems to produce a process of decay similar to that experienced by the corpse.

Andeans protected themselves against this contagion as they did against illnesses in general, through ritual cleansing in a river. Arriaga observes that "the clan comes together with the relatives ten days after a death occurs to accompany the nearest relative to a spring or flowing stream that has been agreed upon. There they duck him three times and wash all the dead person's clothes."[35] A modern account from northwestern Argentina describes a similar ceremony occurring nine days after a death. "Everyone goes to the nearest stream or river, taking the saddled horse, dog, and all of the possessions of the deceased with them. . . . Once there, the friends of the widow wash and bathe her, and when her hair has been well washed, they comb it. When this operation is over they go on to carefully wash all of the possessions."[36]

The flesh of a corpse is conceptualized as wet and "chaotic" in modern Andean thought, while the bones are considered to be dry and "structured."[37] In the modern province of Cuzco the soul (*alma*) of a bad person is thought to be tied to the rotting flesh of its corpse and condemned to wander about causing harm to the living, while the soul of a good person resides in the dry bones of the corpse and exerts a beneficial influence.[38] While a Catholic influence is evident here, the dichotomy posited between the wet flesh and the dry bones of the corpse is most likely derived from pre-Columbian theories of the body.[39]

The underworld, Ucu Pacha, was the symbolic location both of the dead and of the dark, fluid future in Andean cosmology. Thus the dead, through their connection with Ucu Pacha, also participated in the future, which made them particularly useful as oracles to the Andeans. (In the Andes, as has been noted, time was thought to be cyclical, and thus the past was believed to return again, with variations, in the future.) The use of the word *mallqui* (young plant) to refer to the body of an ancestor conveys the concept of the ancestor's involvement with the future. Like plants, the bodies of ancestors mediated between the world underground and the world above ground. This mediating role is perhaps one reason why in the Cuzco region the soul that remains with the body after death is called the middle soul.

The authority of the mallquis was such that an ayllu, or community, could find itself in the power of another people if the latter gained possession of the body of its founding ancestor.[40] Most of the local huacas the Incas brought to Cuzco from the regions they conquered likely formed part of the ancestor cults of their native communities. By taking control of his subjects' ancestors, the Inca could establish himself as the symbolic father of his people.

Removing mallquis from their places of origin, however, disturbed their relationship with their land, for after death the body remained clearly allied to the land from which it had come and out of which, according to Andean cosmology, it was made. Modern Andeans believe it is necessary to be buried in one's homeland in order to preserve one's ties to one's land and community.[41] The same belief may well have been what prompted the Incas to bury their favorite wives with earth brought from their native communities.[42] Through such a practice the wife's homeland would also be symbolically integrated into the sacred Inca capital.

The mummies of the Incas were the most important mallquis in Tahuantinsuyu. As the Inca was divine, and as he represented the empire, it was important that when he died his body be properly cared for and not allowed to decay. It is possible, as well, that the Incas believed that the bodies of the dead would be resurrected one day.[43] For these reasons the corpses of the Incas were skillfully embalmed so that they were able to last for centuries.

The Inca mummies were usually kept wrapped in layers of cloths with their faces covered. Garcilaso de la Vega, however, had a chance to see five of these royal mummies, three male and two female, without their customary wrappings, in 1559 when they had been appropriated by the Spanish. He writes that "the bodies were so well preserved that they lacked neither hair, eyebrows, nor eyelashes. They were wearing clothes as they would in life, with *llautos* [headbands] on their heads, but no other royal insignia. They were seated in the Indian fashion and had their hands crossed on their chests, the right over the left, and their eyes cast downward."[44]

The perfect preservation of the Inca mummies, which were drained of body fluids, as opposed to the putrefaction of ordinary, untreated corpses, served to exemplify the structure and permanence of the Inca state.[45] As well, the dry, stable condition of the mummies made them relatively safe for living humans to be in contact with.[46]

Each mummy was accompanied by its huaque (brother or double) which was a statue chosen in life by the Inca to represent him. A huaque might or might not resemble its Inca brother and was made of gold, stone, or some other material. Each huaque had its own estate and retinue and was treated with almost the same reverence as its human counterpart. The huaques continued to be maintained and revered after the death of their brothers in the same way as the mummies of the latter were. The practice of taking a huaque was not peculiar to the Incas but was relatively common among the Andeans.[47]

In Inca and Andean mythology there are a great many instances of humans who were turned into stone. Manco Capac's brother, who became the stone huaca on the top of Huanacauri hill, is one example of this. Another is Manco Capac himself, who was said to have turned to stone at death, for which reason he had no mummy.[48] The reasoning behind the huaque, which literally means one-half of a pair, would seem to be that an organic body and inorganic stone could not only be fused into one, they could also be maintained separately as two halves of a fundamental pair. The inorganic huaque would thus complement the organic body of its double, providing the permanence and stability the latter lacked.

According to Inca tradition, the cult of the royal mummies was instituted in its full form by Inca Pachacuti at the time when Tahuantinsuyu was assuming imperial proportions. He is said to have disinterred the bodies of the previous Incas, adorned them with gold, and arranged for festivities to be held with representations of the life of each one.[49] Pachacuti evidently recognized the value of the bodies of his royal predecessors as imperial symbols.

There is some disagreement in the chronicles as to where the mummies of the Incas were ordinarily kept. Garcilaso de la Vega writes that they were kept in the temple of the Sun, Coricancha; the male mummies with the image of the Sun and the female ones with that of the Moon.[50] Other accounts place each mummy in its own palace, and indeed, the mummies were discovered by the conquistadores in different places in and around Cuzco.[51] Cobo reconciles this difference by saying that at first the mummies were all kept in the temple of the Sun, but later each one was taken charge of by its descendants and brought to the temple only on special occasions.[52]

The Inca retained all of his land and property after death and these were placed in the care of his panaca, consisting of all his descendants in the male line except for his successor, which maintained them on behalf of his mummy.[53] The possession of the body of its imperial ancestor endowed a panaca with not only a generous living derived from the Inca's property,[54] but also a significant amount of power, for the mummies were often consulted through their attendants on matters of imperial importance.

Huayna Capac, the eleventh Inca, for instance, asked the mummy of his father, Tupac Inca Yupanqui, for permission to marry one of his (Huayna Capac's) sisters who would not agree to the match. The mummy refused to respond and instead there were ominous signs in the sky, so Huayna Capac was obliged to abandon the idea and marry another sister.[55] Huascar, the son of Huayna Capac, also asked the mummy of Tupac Inca Yupanqui for permission to marry; however, in this case "the persons who represented the body . . . consented in its name."[56]

As repositories of authority and tradition, the royal mummies represented the Sun and structure.[57] As sacred mediators, however, they were more closely associated with rain and fluidity. The use of the term *yllapa* (thunder and lightning), to refer to the Inca dead brings out the mummies' role as mediators of the divine and their association with rain. This connection with rain is further brought out by a drawing by Guaman Poma in which a skull labeled "yllapa" is positioned above a rainlike stream of chicha being poured into a jar (fig. 11). The mummy of Inca Roca, who reputedly provided Cuzco with its water supply,[58] was apparently thought to have a special ability to attract rain. Cobo relates that, "whenever there was a need of water for the crops, [Inca Roca's mummy] was carried out in procession through the fields and highlands, and this was believed to be in large measure responsible for bringing rain."[59]

The Inca mummies, like other mallquis, served as sources of information about the future. Cieza de León writes that once a year the Incas gathered all the mummies and sacred statues together in the plaza of Cuzco and asked them one by one through their attendants about the events of the coming year; principally how the crops, the Inca, and the empire would fare: "And if the Incas did not do this every year, they went about fearful and uneasy and didn't hold their lives safe."[60] By serving as conduits between the future and the present, the mummies helped to channel the fluidity of the future and render it manageable.

Figure 11. Inca Funerary Rites
(From Felipe Guaman Poma de Ayala, *El primer nueva corónica y buen gobierno,* vol. 1, ed. J. Murra and R. Adorno [México: Siglo Veintiuno, 1980], p. 262.)

On all important occasions the mummies were brought out from their palaces to the plaza of Cuzco.

> In front of the dead they lit a fire of a certain kind of wood that had been carved and cut very evenly. In it they burned the food that had been placed for the dead bodies to eat. . . . They also had some large cups like pitchers, called *vilques*, made of gold and silver, in front of the dead and in these they poured the chicha they offered to the dead after first showing it to them. The dead would toast each other, and toast the living, and vice versa through their ministers who acted on their behalf. When these vilques were full, they were emptied into a round stone kept as an idol in the middle of the plaza. Around the stone there was a small cistern through which the chicha ran down along certain drains and hidden pipes.[61]

On lesser occasions, similar rituals were performed with the mummies' huaques.

It can be seen here that important Inca ceremonies required that the ideals of equilibrium and exchange be reasserted among the dead and also between the dead and the living, and the dead and the different levels of the cosmos. The dead engaged in reciprocal toasts with each other and with the living, while the food that was burned for the dead traveled upward to the dry Upper World, and the drink that was offered them traveled downward to the fluid Lower World. Categories, thereby, were carefully maintained separately, but also carefully integrated.

These rites practiced by the Incas were based on long-standing Andean tradition. Where the Incas differed from general practice was in the elaboration of the cult of the dead to imperial proportions. The greater the Inca empire became, the more important it was that the bodies of the Incas, which symbolized the empire, should reflect that greatness. The consequence of this, however, was that a large amount of the empire's resources ended up being used for the cult of the dead, and that each Inca had therefore to start anew in collecting land and property for his personal use and postmortem cult.

According to Pedro Pizarro, the Inca Huascar declared, shortly before the arrival of the Spanish, that the dead should be buried and have their riches taken away because they had all that was best in the kingdom.[62] The cult of the dead was too firmly entrenched in Inca tradition and too profitable for the panacas of the royal mummies, however, for such a deposition to be possible. What happened instead was that Huascar's half brother Atahualpa, with the panacas behind him, challenged Huascar's right to the throne and defeated him in battle.

Ironically, while striving to maintain the sacred and social structures of their empire through the worship of their dead, the Incas ended up creating an economic imbalance that led to civil war and left their state riven and prey to the Spanish conquerors.[63]

7
THE METAPHORICAL BODY

THE IMPERIAL BODY

The fundamental divisions of Inca culture—right/left, high/low, external/internal, male/female—had their origin in the human body, as was discussed in Chapter 2. Body parts, and the body as a whole, also provided organizing symbols for the Incas. Garcilaso de la Vega, for example, compares Hanan and Hurin Cuzco to the right and left arms of a body.[1] Each of these "arms," in turn, was divided into five panacas, which could be thought of as five fingers.[2] The empire itself was divided into Hanan and Hurin halves, which likely also carried the metaphorical connotation of halves of a body,[3] while the four quarters of Tahuantinsuyu would correspond to the four quarters of the body.

These four quarters of the empire (fig. 12), in turn, appear to have been symbolically embodied by the Incas in the pirac ritual in which the faces of the participants were divided into four quarters by a line of blood painted from ear to ear. Zuidema finds a similar symbolism occuring in another ritual in which people painted their noses yellow and their faces red in order to communicate with the sun.

> The nose in this case symbolized a gnomon, measuring the passage of the sun through its zenith, and . . . as such it was an *axis mundi.* In Cuzco, a tall round building, the *Suntur huasi*, served that purpose. The eyes of the painted face symbolized the two ritual places on opposite sides of the gnomon. They were used, together with the gnomon, for defining sunrise and sunset positions in relation to the zenith passage of the sun at noon. Two such ritual places in Cuzco, the *Ushnus*, were also used as basins leading to the Underworld for libations to the Sun and to the ancestors. We can conclude then, that whereas the colour yellow is referring to the sun, the eyes with the blood should be conceived as entrances to the Underworld.[4]

In rituals such as these the participants, through metaphor, incorporated the ideal structures of their culture into their own persons.

The preeminent metaphor for the body of the empire was the body of the Inca. We have seen in Chapter 6 how the health of the Inca was intrinsically linked with the health of the empire, so much so that when the Inca was ill his subjects would perform the neces-

Figure 12. The Four Quarters of Tahuantinsuyu (above) with Cuzco in the Center, and the Four Quarters of Spain (below) with Castilla in the Center
(From Felipe Guaman Poma de Ayala, *El primer nueva corónica y buen gobierno*, vol. 1, ed. J. Murra and R. Adorno [México: Siglo Veintiuno, 1980], p. 34.)

sary health-restoring rites of confession and penance. The myth describing how Inca Roca found a source of water, in turn, depicts a metaphorical relationship between the body of the Inca and the land. The Inca's earache is equivalent to the drought suffered by Cuzco, and the blood flowing from his ear corresponds to the underground river he discovers.

While the Inca represented the whole of the empire, at the same time he symbolized its apex (head, or uma) and its center (heart, or sonco). The position of the Inca as the head of the social body is brought out most clearly by Betanzos: "[Inca Pachacuti] named the whole of [Cuzco] Lion's Body, saying that its neighbors and inhabitants were like the members of the body and that he was its head."[5]

Murúa writes that the Inca was carried on his litter by four lords who represented the four quarters of the empire.[6] The body of each lord here evidently represented those of the inhabitants of his particular quarter of the empire, while that of the Inca, carried above and in the center, represented the dominant Inca ideal. Murúa states, in fact, that this practice was intended to keep the inhabitants of the four quarters subordinate to the Incas.[7]

The Inca's pivotal position is set forth in the myth of the origin of the Incas in which Manco Capac thrusts the staff of the Sun into the ground, uniting the sky and the Earth and establishing himself as an axis mundi. The Inca and his wife, the Coya, as the founders of civilization, become the models for all humanity.

In accordance with this ideal, those in the service of the Inca, and the inhabitants of Cuzco in general, were expected to serve as models in their various capacities. Garcilaso de la Vega comments that "the attendants in the service of the royal house . . . did not undertake their functions as individuals. Certain villages, rather, undertook to supply persons for each office. . . . Carelessness or negligence on the part of any one of these servants was held against the whole village, and for the fault of the individual all the inhabitants were punished more or less severely, according to the crime."[8]

CUZCO AS A BODY

As the Inca represented both the whole of the empire and its center, so did Cuzco. Cuzco was divided into four parts by four principal streets that met in the central square. Outside the city these streets became the four highways of the empire. The correspondence between Cuzco and Tahuantinsuyu is made explicit by Garcilaso de la Vega in a chapter titled "The City Contained the Description of the Whole Empire."

> Manco Capac ordered that the savages he had subjugated should settle [in Cuzco] according to the places they came from: those from the east to the east and those from the west to the west, and so on. The chiefs . . . built their houses according to the locations of their provinces: If [a chief's province] was to the right of his neighbor's, he built his house to the right; if it was to the left, then to the left; and if it was behind, then behind. This was undertaken in so orderly a fashion that whoever looked carefully at those neighborhoods and the houses of the many and diverse peoples who inhabited them would see the whole of the empire at once, as though in a mirror or in a cosmographic drawing.[9]

While this description probably depicts the ideal more than the actual, the quadripartition of Cuzco did indeed parallel that of the empire, with each quarter of the city corresponding to a quarter of Tahuantinsuyu. Through controlling this microcosmos, the Incas hoped to control the macrocosmos and establish an ordered and integrated world.

As the center of Tahuantinsuyu, Cuzco provided a fixed point of orientation around which an ordered world could be established. References to Cuzco as the heart and head of Tahuantinsuyu abound in the chronicles. As "heart" and "head" were common synonyms in Spanish for a capital city, most of these references could simply be regarded as reflecting Spanish usage. Fernando de Montesinos, however, gives two instances where these terms appear to occur within an indigenous context. "As Inti Capac decided that the king should be like the heart, in the middle of his kingdom, he ordered that the royal court should be in Cuzco, as it was in the middle [of the kingdom]."[10] Later, he has Inca Roca refer to Cuzco as the head of the empire.[11] Guaman Poma also consistently refers to Cuzco as the head of the kingdom. When describing the four quarters of Tahuantinsuyu, he states that Chinchaysuyo is on the right hand, Collasuyo on the left hand, and Cuzco is the head, in the center.[12]

Garcilaso de la Vega writes that the meaning of Cuzco for the Incas was "navel of the land," and that "navel was a fitting name for it, because Peru is long and thin like a human body, and [Cuzco] is almost in the middle."[13] As navel, Cuzco was in the center both of the vertical body of the cosmos and the horizontal body of the land. It provided an umbilical cord between Heaven and Earth, and it concentrated and radiated power and meaning throughout Tahuantinsuyu.

An example of Cuzco's dynamic role in integrating the empire is found in the capacocha rite. In this rite human sacrificial victims were gathered in Cuzco from the four quarters of the empire. Some of these victims would then be sent home to be sacrificed there, thus carrying sacred energy from the center to the four quarters of Tahuantinsuyu. In his article on this ritual, Pierre Duviols notes that "the trajectory of sacred blood from the Heart (Cusco) through the circulatory system . . . gave life to the head and the brain (the Inca) at the same time as it irrigated and dyed one colour the whole body of the national territory."[14]

If Cuzco was the sonco, or heart and entrails, of the body of Tahuantinsuyu, then the sonco of Cuzco was Coricancha, the temple of the Sun. Betanzos, in fact, describes the temple of the Sun as the dividing mark between upper and lower Cuzco.[15] Within the world at large, therefore, Tahuantinsuyu was sacred and what lay outside profane; within Tahuantinsuyu, Cuzco was sacred and the rest of the empire profane; within Cuzco, Coricancha was sacred and the rest of the city profane.

As the innermost sonco of the empire and the cosmos, Coricancha (Golden Enclosure) stood for both the empire and the cosmos, just as in Quechua the sonco of a person can metaphorically stand for the whole person. The temple, in fact, is described as containing within itself models in gold and silver of all the important elements of Inca life,[16] as well as a diagram of the cosmos (see fig. 1). Thus, Coricancha safeguarded the integrity of the bodies of the empire and the cosmos.

We have already seen how close the correspondence was between Cuzco and the person of the Inca. Cuzco, in fact, was the Inca's court, so that, in a real sense, wherever the Inca was, Cuzco was. When the tenth Inca, Tupac, was involved in a lengthy war far from the capital, he had a replica of Cuzco built in which to dwell, which he named New Cuzco.[17] Huayna Capac, the eleventh Inca, remodeled the city of Quito, where he lived for an extended period, after Cuzco.[18]

In consideration of the close relationship between the Inca and the capital, Inca Pachacuti's description of himself and his court as forming the body of a lion (or puma, the New World lion), logically also applied to the city of Cuzco itself. In fact, this puma metaphor was used by Pachacuti after he had finished redesigning Cuzco according to the clay model he made of the city. Significantly, it was immediately after Pachacuti had assembled Cuzco as a body that he was asked to be Inca.[19]

As discussed in Chapter 4, this act of recreating Cuzco, coming, as it does, at the opening of a new age for the Incas, repeats the original act of Creation. Whereas Viracocha modeled human bodies out of clay and then named them, however, Pachacuti models the social body out of clay and then renames it.

The correspondence between the social and physical bodies of Cuzco is made clear by Pachacuti's restructuring of the city. He assigns that part of the city above Coricancha, Hanan Cuzco, to pure-blooded Incas, and that part below the temple, Hurin Cuzco, to Incas of mixed blood. The very end of Hurin Cuzco, where the two rivers running through the city meet, he is said to have called Pumachupan (Puma's tail).[20] Sarmiento de Gamboa, in his history of the Incas, writes that Pachacuti's son, Tupac, "remembered that his father Pachacuti had called the city of Cuzco the Lion City and that the tail was where the two rivers that ran through the city met, and that [Pachacuti] had said that the body was the plaza and the surrounding settlements, but that the head was missing; however, one of his sons would put it on."[21] The head given to the city by Tupac turns out to be a fortress, known as Sacsayhuaman (Royal Eagle), situated on the hill at the upper end of Cuzco.

Zuidema, in his analysis of this tradition, believes that Sarmiento simply took the lion's body passage from Betanzos's earlier work and rewrote it for his history. "The fanciful nature of Sarmiento's reinterpretation is clear. He wants us to think that Pachacuti Inca had a city without a head, and that Pachacuti Inca himself was not the 'head' at all. Worse, Sarmiento adds a head, the fortress of Sacsayhuaman, the name of which means Royal Eagle (González Holguín). Thus, we have a puma's body with an eagle's head."[22]

The fact that Sarmiento's description of Cuzco as a city without a head appears discordant with the tradition presented by Betanzos in which Pachacuti describes himself as the head of Cuzco suggests, however, that Sarmiento's material may originate from another source than Betanzos. It is also possible that, as it is the physical body of Cuzco that is referred to in Sarmiento's passage rather than the social body, the former might be thought to lack a head while the latter did not.

There is no reason why the fortress of Sacsayhuaman could not have been thought of as an eagle within one context and a puma's head within another, just as the Inca Pachacuti could symbolize Illapa (Lightning) in one context and Inti (the Sun) in another. As well, the practice of creating mythical beasts out of elements taken from different animals was not unknown to the Incas. Pachacuti Yamqui, for example, describes an enormous snake with teeth, ears, and a beard that was said to have appeared at the birth of one of Inca Pachacuti's sons.[23]

We know that the Incas called the lower end of Cuzco, where the two rivers met, "puma's tail." It would therefore be quite natural for them to think of the hill and fortress at the other end of their city as the puma's head. Why, however, would the body of a puma be chosen as the metaphor for Cuzco rather than the human body? The reasons for this lie in the symbolic meaning of the puma for the Incas.

The puma represented a sacred double for the human body in Inca ritual. Cieza de León writes that Inca Pachacuti made his declaration of war against the Chancas "with a lion skin over his head so that it would be understood that he was going to be as strong as that animal is."[24] Molina of Cuzco says that an image of the Sun appeared to Pachacuti, prior to the war with the Chancas, as a man with a puma over his shoulders, another between his legs, and a snake running up his back (another instance of different animals combined in one body).[25] In the huarochico rite the Inca elders greeted the newly initiated boys dressed in puma skins, thus revealing their sacred puma identity at the moment of the boys' entrance into adult society.[26]

We can see from these instances that the puma was ritually employed by the Incas as an "other" for the human body at important times of transition, such as before war and in the male initiation rite. Zuidema deduces from this data that the Incas associated the puma with transition and transformation.[27] Although certainly used at times of transition, the puma seems more specifically to have had the role of affirming Inca identity and power in liminal situations.

This may be why a puma's body was kept in a shrine in a mountain pass that marked the place where one lost sight of Cuzco,[28] and why, according to Garcilaso de la Vega, pumas were kept at the lower end of Cuzco.[29] Enclosing all of Cuzco within a symbolic puma's body would serve to sanctify the whole city and protect it from the forces of chaos outside.[30] The choice of the puma as a metaphor for Cuzco may otherwise have arisen quite naturally from the likeness of the tawny hills of Cuzco in the winter to the body of a puma.

With regard to the puma as a symbol for the center, it is interesting to note that in one of Guaman Poma's illustrations, in which an Andean is depicted surrounded by wild beasts symbolizing colonial administrators (fig. 13), a speckled dragon is at the Andean's head, a tawny puma at his chest, and a plain-coated mouse at his feet.

Figure 13. An Andean Preyed on by the "Wild Beasts" of Post-Conquest Society (From Felipe Guaman Poma de Ayala, *El primer nueva corónica y buen gobierno*, vol. 2, ed. J. Murra and R. Adorno [México: Siglo Veintiuno, 1980], p. 655.)

(The speckled dragon is opposed by a speckled jaguar attacking the Andean behind his back, the tawny puma by a tawny fox, and the plain-colored mouse by a plain-colored cat.)[31] Likewise, in a modern Andean folktale from southern Peru a corpse is reanimated by a speckled cat entering his head; a yellow cat, his heart; and a white cat, his feet.[32] These examples point to a definite association between the puma and the center, or sonco, of the human body.

Another organic metaphor applied by the Incas to Cuzco was that of a quinoa plant. The Incas, in fact, worshipped a large quinoa root from which they held Cuzco had sprung, and which they believed continued to sustain the city.[33] This root would have provided Cuzco with a direct conduit to the Lower World, while the city itself, as a "tree," reached up into the Upper World. The growth of this metaphorical tree can be seen as the direct result of the supernatural fertilization of the Earth that occurred when the golden staff of the Sun was plunged into the ground at the future site of Cuzco by Manco Capac.

This understanding of Cuzco as a plant would not contradict the body metaphor, as plants and bodies were often associated by the Incas. One myth from the Huarochiri region, in fact, describes humans as having originated from the quinoa plant.[34] The most common instance of an association between plants and bodies is that which occurs in the use of the term *mallqui,* which can mean both "ancestor" and "young plant." Interestingly, in the modern Cuzco region pumas are also sometimes referred to as "ancestral co-parents" (*machu compadre* and *paya comadre*).[35] As the origin of civilization for the Incas, Cuzco would also have had the role of an ancestral parent, helping to guide the future of the empire, while as a human city, a puma, and a tree, it would have possessed the combined vitality of all the living things on Earth.

John H. Rowe states that Cuzco was actually laid out in the form of a body of a puma as seen from the side (fig. 14), with its outline described by streets, buildings, and geographical features.[36] This argument is also taken up in an urban study of the Inca city published in 1980, which holds that the Incas planned to incorporate future expansion of the city into the puma's body by changing the outline of the puma so that it would appear to be crouching.[37]

Zuidema, however, considers that the passages from Betanzos and Sarmiento in which the puma metaphor is applied to Cuzco, and on which Rowe and others largely found their case for Cuzco's having the form of a puma, are too heavily influenced by European

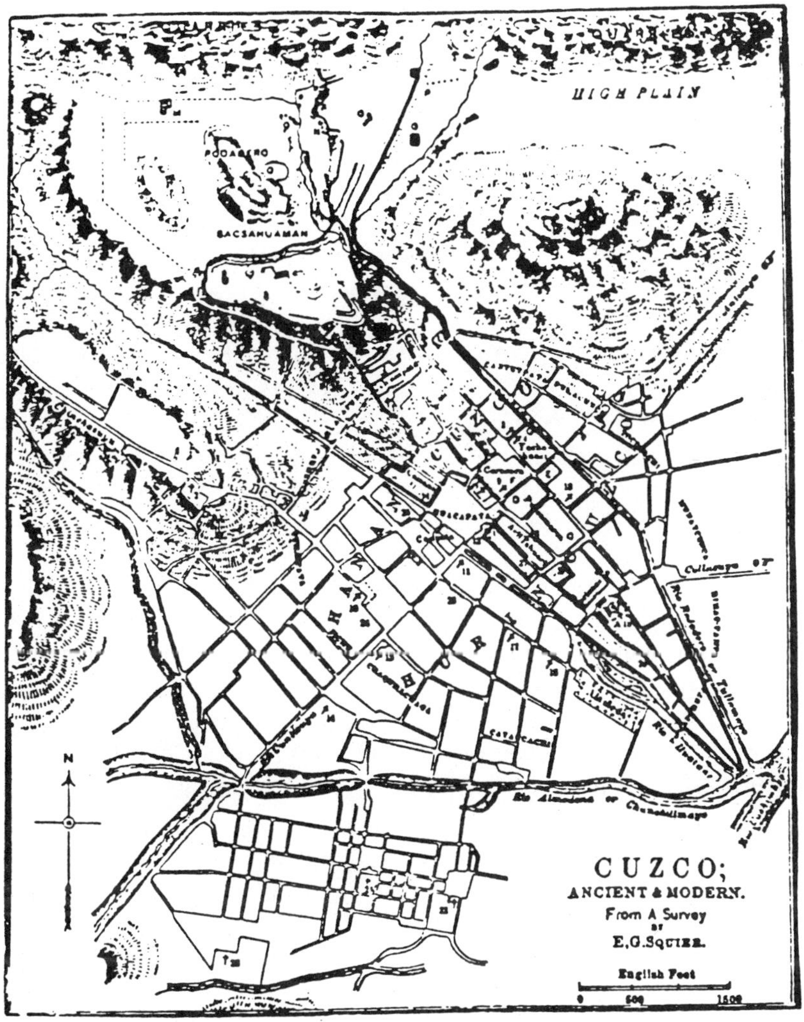

Figure 14. Cuzco as a Puma
(From John H. Rowe, "What Kind of Settlement Was Inca Cuzco?" (Ñawpa Pacha 5 [1967]: p. 77.)

concepts of the body politic and the mystical body of Christ to be reliable.[38] That the Spanish chroniclers may have understood the body metaphor used in the Andes in terms of their own cultural background, however, does not prevent the metaphor from being authentically Andean in origin.

The cyclical structure of Inca tradition, according to which Pachacuti's creation of Cuzco is a repetition of the original act of

Creation, makes it logical that Cuzco, which takes the place of human bodies in this second creation, would have the symbolic meaning of a body. Indeed, the fact that the Incas believed that Cuzco had sprung from a root demonstrates that the city was conceived of by them as an organic body, and not simply an ensemble of structural oppositions.

Modern Andeans continue to order and understand their society through the use of body metaphors. In the Bolivian village of Chucuito, for example, the inhabitants of the upper half were traditionally hatmakers and those of the lower half shoemakers, suggesting an association of the former with the head of a body and the latter with the feet.[39] In the region of Jesús de Machaca in Bolivia the metaphor of a body is used to structure and integrate communal life and land. The communities forming a common body are conceptualized as body parts: head; shoulder; trunk; or foot. The higher its position in the metaphorical body, the greater a community's status.[40]

Such metaphors are so firmly rooted in Andean culture that it is unlikely that they are solely the result of the influence of Western political and religious models. While Cuzco may not have actually been designed in the form of a body, therefore, it is safe to conclude that the metaphor of the body was applied to it by the Incas and used to convey a complex series of concepts including centrality and wholeness, symmetry and hierarchy, animation and exchange.

This discussion of Andean body metaphors begs the question of whether the Incas themselves made use of terms similar to "metaphor," "symbol," or "model," or whether these are simply a convenient way of apprehending Inca practice through Western categories. While we do not have enough data to answer this question properly, there is evidence to suggest that such concepts would not have been entirely foreign to the Incas.

In González Holguín, for example, we find the Quechua term *ñirac*, meaning "such" or "similitude," as in the expression *nocañirac* (such as I). The word *unancha*, in turn, meant "sign" or "insignia," such as one might find on a coat of arms. The verb *unanchani* meant both "to make signs" and "to understand." An unancha, for the Incas, therefore, was a sign that revealed meaning, an index, rather than a symbol that required deciphering. The latter connotation appears to be more closely signified by the Quechua word *quellca*, which meant "drawing," "painting," and, after the Conquest, "writing." There can be little doubt that the drawings and designs of the Incas encoded a great deal of data and that the term *quellca* therefore implied a form of visual symbolism.

The Quechua term *hamu,* on the other hand, signified "mode," "style," and "type." The verb *hamutani* meant "to make a model," "to copy," "to organize," and "to reason out," while the noun *hamuyniy* meant judgment and will. This suggests that when the Incas created models of their world, as, for example, they did in Cuzco, they were at once engaged in an act of organizing the world and of understanding it.

As we shall see in the next section, the Incas understood the "metaphors" and "models" that ordered the cosmos in a much more concrete and active way than we in the West usually do. That is, the Earth, for example, was not simply "like a body" for the Incas, but actually a body, with a life and will of its own.[41]

THE COSMIC BODY

A myth from the central coast of Peru, recorded in the seventeenth century, tells how Pachacamac (Animator of the World) killed his half brother and then scattered the pieces of his body over the land in order to create food. From the teeth corn grew, from the bones grew all tubers, and from the flesh grew all other vegetables and fruits.[42]

Understanding the land in terms of a body that provides one with food (and that, in turn, has to be fed through ritual) is a common practice in the Andes. The most extensive examination of the phenomenon was undertaken by Bastien in a community of Qollahuaya Indians on Mount Kaata, in midwestern Bolivia.

The Qollahuayas studied by Bastien conceptualize the mountain on which they live in terms of a body: The peak is the head, the central slopes are the heart and stomach, and the lower slopes the legs. This metaphor is also applied to the three villages on the mountain: The upper village, Apacheta, corresponds to the head of the mountain body; the middle village, Kaata, corresponds to the heart; and the lower village, Niñokorin, corresponds to the legs (fig. 15).

All three villages participate together in major rituals, each village contributing products characteristic of its particular altitude. It is the special task of the ritualists living in the heart, Kaata, however, to circulate energy throughout the whole of the mountain body by making ritual offerings to the mountain. In the words of one of

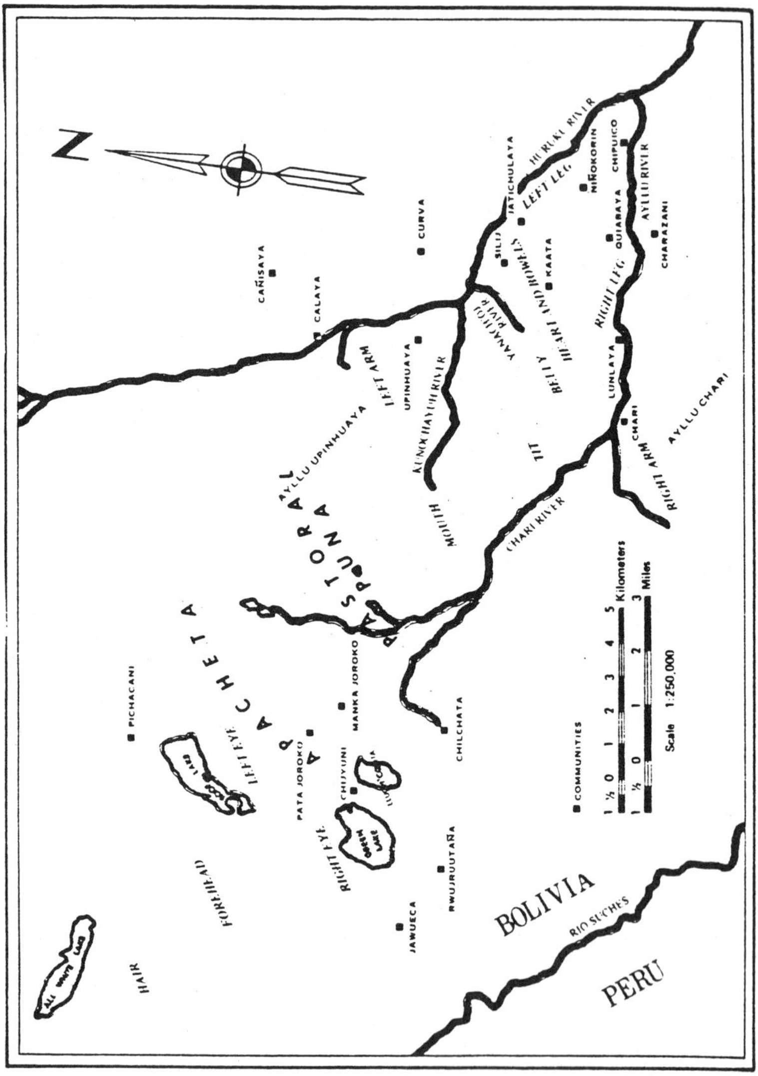

Figure 15. The Body-Land Metaphor of the Qollahuayas of Kaata, Bolivia (From Joseph Bastien, *Mountain of the Condor* [St. Paul, Minn.: West, 1978], p. 44.)

these ritualists, "the mountain is like us, and we're like it. . . . If we don't feed the mountain, it won't feed us."[43]

The three communities on Mount Kaata are united not only through intermarriage and the exchange of goods and services, but through belonging to one body. This sense of collective wholeness has led to centuries of legal battles to prevent outsiders from incorporating land the Quollahuayas consider to belong to the body of Mount Kaata. In the late eighteenth century an eighty-year-old inhabitant of Mount Kaata, protesting the appropriation of Niñokorin by Spaniards, testified to a representative of the Spanish Crown that Niñokorin belonged to Mount Kaata "because it is the leg of its body."[44] The mountain-body metaphor has enabled the inhabitants of Mount Kaata to maintain their identity and solidarity, despite the pressures of external forces, to the present day.[45]

While one particular mountain or area of land may be thought of as a body by the Andeans, on a larger scale the Earth itself forms one body, called Pachamama (Mother Earth). According to one modern Peruvian Andean, "Peru begins in Lake Titicaca, which is the sex of our Mother Earth, and ends in Quito, which is her forehead. They say that Lima is her mouth and Cuzco her beating heart. Her veins are the rivers. But Pachamama extends much farther. Her right hand is Spain, perhaps."[46] An inhabitant of the town of Pinchimuri in Peru says of Pachamama that "we live and work on her. She nurses us and raises us like our mother. . . . She has bones and blood. She has hair too. The grass is her hair. Her blood is in the ground. She always bleeds when she is plowed. . . . We give her seeds and she gives birth. . . . We give offerings to Pachamama for our produce and for our animals, so that the animals don't become sick and so that the harvest will be good."[47]

As discussed in Chapter 5, the Incas also understood the Earth to be a female body, which was violated through agriculture in order to produce crops, but which also had to be fed through ritual offerings and placated by worship. Guaman Poma reports, for example, that diviners sleeping in caves, symbolic of the mouth of the Earth in the Andes, would say, "Mother Cave, don't eat me," and then feed the cave with corn and coca.[48]

The Incas interpreted movements of the body of the Earth as signs, just as they did the twitchings and tremblings of the human body. Murúa reported that, "when the earth trembles, they throw water on her, saying that the huacas are thirsty and want to drink . . . [They say that the Creator] is angry and that he wants to reverse the world, making the high, low, which [event] is called Pachacuti."[49]

The most telling evidence of the metaphorical relationship between the body and the land is found in the language of the Incas itself, Quechua. González Holguín, in his dictionary, for example, states that uma means both "head" and "mountain peak"; *cenca* means both "nose" and "hilltop"; *chaca,* both "hip" and "bridge"; and *ciqui,* both "buttocks" and "mountain slope" (see the appendix). Louisa Stark finds that in modern Quechua *kunka* (neck) can mean "a small mountain chasm"; *ñawi* (eye) "a spring"; *rixra* (shoulder) "a mountain shoulder"; *muqu* (knee) "a hill"; *wixsa* (stomach) "a small hill"; and *siki* (buttocks) "a foothill."[50]

In complementary opposition to the female body of the Earth, the sky represented a male body, although this latter metaphor was not as elaborated as the former in Andean cosmology. Just as there could be many metaphorical bodies within the body of the earth, the sky also contained many metaphorical bodies. The Incas believed that the sun and the moon served as the celestial models for humans, while different stars represented and watched over different animals.[51] The Incas united the bodies of Tahuantinsuyu and Cuzco with these celestial bodies by aligning their sacred edifices with the movements of the spheres.

As the representative of the human social body, situated in between the earth and the sky, the Inca had the task of maintaining the three levels of the cosmos integrated. This task was begun by the Creator, Viracocha, and then taken over by the first Inca, Manco Capac, when he thrust the staff of the Sun into the Earth. We have already discussed in previous chapters some of the rituals employed by the Incas to integrate the cosmos. Three Inca institutions that were used to this effect were the temple of the Sun, the palace of the Inca, and the *acllahuasi,* house of the chosen women. The first represented the body of the sky, Hanan Pacha; the second the body of the human world, Cay Pacha; and the third the body of the Earth, Hurin Pacha. These three buildings were erected together in all the principal regions of the empire.

By creating an institution of virgin women (symbolizing the virgin Earth) in order to serve the Sun through his temple,[52] the Inca established himself as the mediator between the Lower and Upper worlds and the medium of cosmic order. He thus continued the original mythical act in which Manco Capac ritually united the Sun with the Earth and ordered the cosmos, making the human world possible. Through the interdependent institutions of the temple, the palace, and the acllahuasi the Incas kept the three levels of the cosmos separate but interrelated. Each of these institutions, like its corresponding cosmic level, was a world and a body in its own right: Together they formed the cosmos.[53]

It is not possible to ascertain to what extent the Incas applied the body metaphor to the cosmos, as the existing data are suggestive rather than conclusive. The tripartite division of the cosmos certainly corresponds to the tripartite division of the body in Inca cosmology. Hanan Pacha is associated with structure, as is the upper body; Cay Pacha is associated with integration, as is the sonco; and Hurin Pacha is associated with fluidity, as is the lower body.

Some of the evidence for the human body as a model of and for the cosmos in Inca cosmology was presented in the analysis of the Inca cosmogony in Chapter 3. Guaman Poma's drawing of the Creator with his feet on the ground and his head in the sky (see fig. 4) illustrates this model nicely. Garcilaso de la Vega compares the cosmos to a body in his explanation of the meaning of the name Pachacamac: "Pachacamac means 'he who gives life [anima, spirit] to the universe,' and in its original full signification it means 'he who does with the universe what the spirit does with the body'."[54] This fits in with the cosmogony in which Viracocha animates the material world with his voice, and therefore with his breath and spirit.

The body-cosmos model appears to have been temporal as well as spatial. The word *pacha*, in fact, used in the names of deities such as Pachacamac and Pachamama, refers to time as well as space. In the Andean communities of Mount Kaata the head of the mountain is associated with birth, and the foot of the mountain with death. "Animals and people originate and return to the head of the mountain. It is the place of origin and return, like the human head which is the point of entry and exit for the inner self. The dead travel by underground waterways to the mountain's head (*uma pacha*) from which they can arise to the land of the living. The living emerge from the eyes of the mountain, journey across its head, chest, trunk, and legs, and die in the lowlands."[55]

We find the term *uma pacha* (head of space/time) also used in the early seventeenth-century Quechua manuscript of the Huarochiri province, in which human sacrificial victims are described as being hung together with offerings of corn and other goods so that they will return to the Uma Pacha, the point of origin.[56]

The head was thus associated with the past (although Western terms such as *past* and *future* apply only loosely here) in the Andes; going to the head of space/time was equivalent to returning to the original structure. The head, in fact, was apparently thought to be the location of the memory in Inca thought, as contrasted with the sonco, the site of reason (see the appendix).

The feet, in turn, were associated with the future, the chaos of the unknown, and with the dead. In the modern Andean belief recorded by Bastien the dead become fluid at the foot of the mountain, and are restructured and reborn at the head in a cyclical fashion.[57] The trunk of the body corresponds to the living and to the present, and mediates between the feet and head (the past and the future).

In the cosmos the head corresponds to the Upper World, the place of original ideal forms represented by the heavenly bodies, and the feet correspond to the Lower World, the place of transition. The feet (future) move the cosmic body while the head (past) remains still. The right-left alternation of walking is suggested by the alternation of day (associated with the right) and night (associated with the left), and the dry (associated with the right) and wet (associated with the left) seasons. Viracocha sets the body of the cosmos in motion with his animating breath; without his continued input it would, in the words of the Inca prayer recorded by Pachacuti Yamqui, "grow tired and die."

As was mentioned in Chapter 2, the Incas also associated the front of the body with the past and with clarity, and the back with the future and with darkness.[58] The ordinary passage of time, which was but a repetition of known patterns, could be represented by the metaphor of walking. A truly revolutionary event, a pachacuti (reversal of space/time), however, could be represented only by the metaphor of the body's turning around and facing the opposite direction (or, on the vertical axis, by being stood on its head). When this occurred the structures of the past would become submerged in the fluidity of the future, and the world would be restructured according to new principles. This is why the Andeans of Huarochiri said of the deity known as Pachakuyuchiq (World Mover) that "when he turns his face [the Earth] moves as well. . . . If he were ever to turn his whole body, the world would end."[59]

8
THE DISORDERED BODY

THE BREAKUP OF THE BODY

By the time of the eleventh Inca, Huayna Capac, Tahuantinsuyu had grown so large—extending from present-day southern Colombia to Chile—and so heterogeneous as to be virtually impossible to govern efficiently. Huayna Capac exacerbated this situation by residing chiefly in the city of Quito in the northern end of the empire, instead of in Cuzco, the capital. This division of power between Quito and Cuzco gave Tahuantinsuyu, to use the word Cieza de León attributes to Inca Tupac, two "heads."[1]

When Huayna Capac died around 1525, his body was embalmed and carried back to Cuzco to be honored in the traditional manner. His heart, however, is said to have been kept in Quito.[2] This separation of the Inca's body between the two centers of the empire foreshadowed the separation of the empire itself.

Apparently in line with Huayna Capac's wishes, Huascar, one of Huayna Capac's sons by the Coya, assumed the leadership of Cuzco and its dominions, while Atahualpa, one of his sons by another wife, took control of Quito and its surrounding territory. The two half brothers shortly went to war with each other for control of the whole empire. After a devastating battle Atahualpa emerged victorious. It was while he was preparing to officially assume the Incaship of all of Tahuantinsuyu that he came into contact with Francisco Pizarro and his band of soldiers in 1532.

The crucial encounter between the Inca and the Spanish took place in an enclosed square in the town of Cajamarca to the north of Cuzco. The priest accompanying the Spanish expedition gave Atahualpa a brief summary of Christian doctrine, denounced Inca religion as invented by the Devil, and demanded that the Inca become a vassal of the Holy Roman emperor. While giving this address, the priest held a book, either a Bible or a breviary, in one hand.

Atahualpa, deeply offended by this speech, which was received by him in a garbled form through an interpreter, demanded of the priest by what authority he made these claims. The friar held the book up to him. Atahualpa examined it, but as it said nothing to him, he dropped it to the ground. This "sacrilege" was the excuse the Spanish needed to attack Atahualpa's retainers and take the Inca prisoner.

Atahualpa, aware of the conquistadores' desire for gold and silver, offered to fill a large chamber with gold and another with silver in return for his freedom. Once this ransom was paid, the highly legalistic Spanish realized that their grounds for continuing to hold the Inca were shaky. They therefore accused him of a series of crimes, tried him, and condemned him to death. Before his execution Atahualpa allowed himself to be baptized in order to attain the privilege of being garroted rather than burned at the stake. The Inca was probably concerned that his corpse be preserved so that it might be able to participate in the afterlife, according to Andean religion.

Significantly, although Atahualpa was garroted, he is remembered in Andean tradition as having had his head cut off by the Spanish (for example, fig. 16). This imagined dismemberment of the Inca's body mirrored the dismemberment of the imperial body that occurred with the execution of the Inca, for the strictly vertical nature of the Inca hierarchy, with all power flowing from the Inca downward, made the loss of the Inca symbolically analogous to a decapitation.[3]

In 1534, half a year after Atahualpa's execution, Manco, a half brother of the late Inca, was installed as Inca by Pizarro, so that the Spanish might rule the empire through him. Manco was subjected to repeated humiliations by the conquistadores, who went so far as to sleep with his wives and urinate on him.[4] According to one of his sons, Titu Cusi Yupanqui, the Inca said to his tormentors, "you have acted not as the Christians and sons of Viracocha you said you were, but as servants of Supay [the Devil], in whose footsteps you follow, doing evil to those who do you good."[5]

Finally, in 1536, the Inca escaped from his captors and led a revolt against them. The first thing Manco did was attempt to recapture Cuzco. Against incredible odds, however, the Spanish were able to retain the city. Manco then fell back to the mountain retreat of Vilcabamba, not far from Cuzco, where he established his headquarters. From there he conducted raids into Spanish-held territory until 1545 when he was murdered by a group of Spanish outcasts who had sought refuge with him.

After a brief and abortive reign by a young son of Manco, the leadership of the Incas passed over to another of Manco's sons, Titu

Figure 16. The Beheading of Atahualpa
(From Felipe Guaman Poma de Ayala, *El primer nueva corónica y buen gobierno*, vol. 2, ed. J. Murra and R. Adorno [México: Siglo Veintiuno, 1980], p. 362.)

Cusi Yupanqui. Titu Cusi continued to uphold Inca independence from Spain, while maintaining intercourse with the Spanish and even allowing himself to be indoctrinated by Augustinian missionaries. In 1571, still resisting Spanish rule, Titu Cusi Yupanqui died of pneumonia.

Yet another of Manco's sons, Tupac Amaru, succeeded Titu Cusi as Inca, and hostilities against the Spanish were renewed. At this point, weary of attempts at negotiations, the viceroy Francisco de Toledo sent out an army against Tupac Amaru and his followers. After a bitter battle the Inca rebels were defeated, and Tupac Amaru was captured and brought to Cuzco where he was publicly beheaded by the Spanish in 1572. With this execution, which must have appeared to the Andeans a conclusive confirmation of the original execution of Atahualpa and the overthrow of Inca rule, the last remnant of the Inca empire was destroyed.[6]

The breakup of the empire was accompanied by a drastic decline in the population of the region due to the harsh treatment and heavy tributes meted out to the Andeans by the Spanish, and the devastation caused among them by epidemics introduced by the foreigners.[7] At the same time the destruction the Andeans experienced in the social body and in their own bodies was enacted on a sacred level by the systematic destruction by the Spanish church of the holy bodies of Andean religion, the huacas.

It had been part of the Incas' plan of conquest to appropriate the major huacas of the peoples they subordinated. The former were taken good care of by the Incas so long as the latter remained good subjects. No such arrangement was possible with the Spanish, however, who simply and relentlessly destroyed anything connected with the indigenous cult.

Church-appointed missionaries, called visitors, traveled from village to village cajoling and ordering the inhabitants to surrender all their huacas to them. Once the huacas of a village had been gathered, they were quickly disposed of. In the words of one of these visitors, Father Arriaga, "everything that is inflammable is burned at once and the rest is broken into pieces."[8] Often, however, even this drastic procedure was not enough. Arriaga writes that "many years ago . . . a Dominican friar . . . traveled through the province burning and removing huacas, and in some places we have found the Indians worshipping either the name of the burned huaca or the parts of the huaca that would not burn. Therefore, great care must be taken to see that what is left over after burning is cast out where it cannot be seen."[9]

The Inca mummies, among the holiest of the realm, were kept hidden from the conquistadores by their guardians, who continued to attend to them according to Inca ritual. The Spanish, however, indefatigably searched the mummies out and took them away, either burning them or displaying them for a time for the curiosity of their countrymen and then burying them in unknown sites. Cobo describes in a few telling sentences the search for and discovery of Inca Viracocha's mummy.

> Gonzalo Pizarro (Francisco Pizarro's brother) looked for it for a long time because a great treasure was reputed to be buried with it, and in order to find it he burned a few Indians, men and women. At last he found it along with a great amount of its estate, which was given to him by [the mummy's guardians]. Pizarro had the mummy burned; however, the Indians of its ayllu gathered the ashes and placed them together with a certain mixture in a small stone jar next to [Viracocha's huaque], which, as it was of stone, was left alone by Gonzalo Pizarro and his men, who took no notice of it."[10]

When the Spanish learned of this "idolatry," however, even the ashes of Viracocha's mummy were sought out and taken away.[11]

The Incas themselves did not fare much better than the mummies of their ancestors. Many of them were killed fighting against the Spanish, others died from the epidemics that ravaged the country at that time. A contingent of the most noble Incas was exiled by the Spanish to Lima where a great number died from grief and illness.[12] Those Incas who had embraced Christianity and the Spanish way of life ended up being assimilated into Spanish society through marriage. The others eventually merged into the Andean population.

In 1780 José Condorcanqui, "Tupac Amaru," a descendant of the last Inca, Tupac Amaru, led a general native uprising against the Spanish oppressors, which many of his followers undoubtedly believed would result in a restoration of Inca rule.[13] After a series of battles Tupac Amaru II was captured in 1781 and sentenced to be drawn and quartered in Cuzco. After the execution his trunk was burned and his limbs and head were exhibited separately in different Andean towns. (Again the Spanish, with their method of execution, were tapping into a deep reservoir of Andean body symbolism.)

The revolt resulted in the enactment by the Spanish of a set of laws designed to abolish the memory of the Incas among the natives and forbidding even the appellation of Andeans by the term *Incas*. In the decades after this revolt opposition to Spanish imperial rule grew among persons of Spanish descent born in America and led to the war of independence from Spain, the establishment of republics in the former Spanish territories, and the modern Andean era.[14]

THE REVERSED BODY

On the one hand, the devastation caused by the Spanish Conquest was experienced by the Incas as a literal and metaphorical breaking up of the human body. On the other hand, however, the reversed order introduced by the Conquest was manifested by the literal and metaphorical reversed position of the human body in the Andes.

Whereas the Incas had previously identified themselves with dominant masculinity and their subject peoples with subordinate femininity, with the Conquest they found themselves in the latter position, while the Spanish occupied the former. The Incas, in fact, conceived of the swords of the Spanish as penises,[15] indicating that they experienced their defeat at the hands of the Spanish as a symbolic sexual violation.

The Spaniards' actual violation of Andean women, many of whom were forced to serve as their concubines, did, in fact, play an important role in the destruction of Andean society by breaking down its traditional social structures and its ethnic integrity.[16] The conquistadores' violation of the acllas struck at the heart of the Incas' sacred order by disrupting the system of cosmic integration and exchange whereby the inviolate *acllas* were offered to the Sun through the medium of the Inca in return for divine favor.

The Incas, who had expressed their domination over the peoples they conquered through the ritual of stepping on the enemy, were now stepped on themselves by the Spanish. This occurred not only in the reversed social order but also in the very architecture of the former empire, for Spanish edifices were constructed on the razed foundations of Inca buildings. In Cuzco, notably, a church was built on the site of Coricancha, and a convent on the site of the house of the acllas.

The Andeans' experience of Spanish rule as a stepping-on of their bodies is graphically illustrated by Guaman Poma in drawings in which he shows a Spaniard placing his foot on the head of a prone Andean (for example, fig. 17). For the Spanish, the act of stepping on an Indian was probably only one way among many of dominating and humiliating the Andeans. For the Incas and their subjects, it was the ultimate sign of conquest.

Figure 17. A Spaniard Stepping on an Andean
(From Felipe Guaman Poma de Ayala, *El primer nueva corónica y buen gobierno*, vol. 2, ed. J. Murra and R. Adorno [México: Siglo Veintiuno, 1980], p. 510.)

Another reversal of the body that occurred at this time is that the upright human body, which had served as a symbol of cosmic order in the Andes, was turned on its head to serve as a symbol of cosmic disorder. In Quechua the term for "upright," *ticnu,* refers both to the standing human body and to the sun at its zenith, while that for "upside-down," *tiksu,* refers both to an upside-down body and to the sun at its nadir. The same expression used for a reversed body, therefore, also conveys the idea of a reversed cosmos.

In the writings of Guaman Poma we repeatedly find the Spanish expression *mundo al revés* (reversed world)[17] used to convey the disorder of post-Conquest society in which Spaniards who came from the lowest ranks in their own society assumed positions of leadership in the Andes, and Andeans who had been mere workers in Inca times were appointed chiefs by the Spanish.[18]

This reversal of order extended even to the family. Guaman Poma describes children in the post-Conquest Andes as dominating, instead of obeying, their parents, which situation he illustrates in a drawing of a son stepping on his supine father (fig. 18).[19] It is enlightening to recall that this image of a son standing over his father occurred in Inca mythology the last time that the world was reversed, when Pachacuti took over the rule of the empire from his father, Inca Viracocha.

One of the rites of transmission of power in the Inca empire was for the heir to the throne to kiss the feet of the reigning Inca.[20] Pachacuti reversed this by having the reigning Inca, Viracocha, kiss *his* feet. The Spanish went much further in that they took power by cutting off the head of the Inca. This represented a break with tradition and a reversal of order much more serious than that occasioned by Pachacuti. What made the matter worse still was that the Spanish conquistadores, unlike Pachacuti, were not even in the same social class as the Inca.

Guaman Poma draws on the Christian metaphor of Christ as the head of the body of the church (Eph. 4:11–16) to explain to the Spanish why Francisco Toledo, who was a mere viceroy, or subordinate member of the body, had no right to sentence to death the Inca Tupac Amaru, who was the "king of Peru," and therefore the head of the body. "Consider that in your body, the feet cannot rule the head, the hands cannot rule the head, and even the heart, though it may be more important, is not worth anything without the head. Likewise, without God, the Christian is worth nothing . . . and without the king, the body is worth nothing. And therefore, whoever is the head is the king and none other."[21]

Figure 18. A Son Stepping on His Father
(From Felipe Guaman Poma de Ayala, *El primer nueva corónica y buen gobierno*, vol. 2, ed. J. Murra and R. Adorno [México: Siglo Veintiuno, 1980], p. 820.)

In the upside-down world of the post-Conquest, however, the "king" was executed by a "commoner," and the "feet" did rule the "head." Many Andeans experienced this reversal of order in their own bodies, for as a punishment, the Spanish would commonly hang Andeans up by their feet.[22] In Inca times this was the punishment meted out to those who seriously disrupted the social order, such as adulterers and thieves (or persons who disobeyed their parents).[23] The Andeans who were hung upside down by the Spanish (and those who saw and identified with them) consequently, although they may have been completely law-abiding according to Inca customs, felt themselves associated with the worst elements of social disorder.

In the post-Conquest world established by the Spanish, therefore, the Incas went from representing the forces of order to representing those of chaos, and from being associated with the sky and the Upper World to being associated with the Earth and the Lower World. From the Andean perspective, the Spanish divinities—God, Jesus Christ, and the saints—played a fundamental part in this overthrow by replacing the Inca divinities as the lords of the Upper World. The Incas, who had relied on divine favor from above for the support of their reign, now found themselves chastised from above by foreign gods.

In one telling passage Guaman Poma describes how the Spanish patron saint, Santiago (Saint James), defeated Manco's attempt to recapture Cuzco from the Spanish. "Lord Santiago came down with a clap of thunder. Like a bolt of lightning he fell from the sky to the Inca's fortress. . . . And when he fell to Earth the Indians were frightened and they said that yllapa, thunder and lightning, had fallen from the sky, *caccha* [thunderclap] of the Christians, in favor of the Christians. They say he came on a white horse . . . with great death and destruction for the Indians, and he destroyed the ring of Indians that Manco Inca had placed around the Spanish."[24] The accompanying illustration depicts Santiago with his foot on the fallen Inca's head in the ritual pose of stepping on the enemy (fig. 19). Thus the Incas were displaced from the sky and relegated to the low, subordinate position of the conquered.

This turning of the world on its head was, in Andean terms, a pachacuti or pachaticra (translated by Guaman Poma as "that which places the world head down"[25]). In the pachacuti, as it is described in Andean myth, a confrontation between nature and culture characteristically takes place. This was true, as well, of the period of the Conquest.

Initially the Incas considered the Spanish to be divine and asso-

Figure 19. Santiago Stepping on the Inca
(From Felipe Guaman Poma de Ayala, *El primer nueva corónica y buen gobierno*, vol. 2, ed. J. Murra and R. Adorno [México: Siglo Veintiuno, 1980], p. 376.)

ciated them with Viracocha, in part because of their seemingly supernatural powers and in part because they were white and bearded, as Viracocha was reputed to be, and they came from the sea, over which Viracocha was said to have disappeared.[26] However, as the Andeans grew more and more disillusioned with the Spanish, many came to think that the conquistadores in fact embodied savagery in contrast to their own state of culture.

The inhumanity of the conquistadores was felt so deeply by the Andeans that they symbolically equated the Spanish with wild animals. Garcilaso de la Vega, for example, describes the Andeans, after having heard how a cross protected a Spaniard from an attack by a jaguar, as appealing to the cross to protect them from the wild animals they perceived the Spanish to be.[27] Guaman Poma actually depicts the Spanish and their native underlings as wild animals feeding on the Andeans (see fig. 13).[28]

These images of animals preying on humans are reminiscent of the reversed order of the first chaotic, sunless age in Andean mythology when a kind of anticulture prevailed in which animals were said to feed on humans.[29] The metaphorical inversion of the world that occurred with the Conquest, in fact, could be understood not only as a reversal of position, of high and low, but also of time, as expressed by the Andean concept of the pachacuti, which literally means "space/time reversal." From this perspective the death of the Inca, symbolically the Sun, and the irrationality of the Spanish rule that followed returned the world to the first age of darkness and anticulture.

Modern Peruvian Andeans, in fact, remember the Conquest as a time when a plague came, blowing a conch-shell trumpet (reminiscent of the giant in Inca mythology who threatens to flood the world by blowing on a trumpet), and introduced a period of chaos. This age of chaos was characterized by bodily disorder and inversion, for parents are held to have eaten their children, and guinea pigs (raised for food) fed on corpses. The modern Andeans are said to be descended from the survivors of that time.[30]

SENSORY DISORDER

The anticulture of post-Conquest society was manifested to the Andeans not only in the reversed social order, but also in the reversed sensory order. The uncontrolled sensuality of the Spanish occasioned both fear and ridicule on the part of the Andeans.[31] At times, however, the Spanish appeared to suffer from a complete perversion of the sensorium: All they could see, hear, touch, taste, or smell were gold and silver. Guaman Poma writes of the conquistadores that "they were like desperate men, stupefied, crazy, senseless with their greed for gold and silver. Sometimes they did not eat with the thought of gold and silver. Sometimes they had a big fiesta, imagining everything they held in their hands was gold and silver. . . . Their desire for gold and silver was such that they didn't fear death. With their greed for gold and silver, they go to hell."[32]

The Andeans at first believed that the Spanish actually ate gold, as this was the only reason they could think of for their insatiable desire for it (fig. 20). (The Spaniards' head-to-toe armor probably aided this impression by suggesting that the Spanish were "men of metal.") In Andean cosmology gold emanated from the sun and represented timeless, ideal structure.[33] As such it was used for the creation of sacred models.[34] That the Spanish should apparently use gold for food, and thus transform structure into process, signified a complete inversion of order for the Andeans.

The "sensory disorder" displayed by the Spanish that made the greatest impression on the Andeans, however, was writing.[35] Although sixteenth-century Spain still retained many characteristics reminiscent of an oral culture, it was a society penetrated by the effects of writing. The Spanish had a heavy, and almost obsessive, reliance on written documents and texts. Lewis Hanke comments that "Spaniards were so accustomed to certifying every action they took that notaries were as indispensable to their expeditions as friars and gunpowder."[36]

The closest equivalent the Incas had had to writing was the quipu, knotted strings on a cord, which served as a mnemonic device.[37] The quipu, however, was a much more sensual medium than writing, engaging to a greater extent touch and rhythm in the tying of the knots, and involving a wide range of colors and patterns (besides being impregnated by the odor of its creator). As well, the quipu was not flat and linear as writing was. Marcia and Robert Ascher write that

> the quipumaker's strings present no surface at all. . . . A group of strings occupy a space that has no definite orientation; as the quipumaker connected strings to each other, the space became defined by the points where the strings were attached. . . . The relative positions of the strings were set by their points of attachment, and it is the relative position, along with the colors and knots, that render[s] the recording meaningful. Essentially then the quipumaker had to have the ability to conceive and execute a recording in three dimensions with color.[38]

Figure 20. The Inca and a Spaniard. "Do you eat gold?" "Yes, we eat gold." (From Felipe Guaman Poma de Ayala, *El primer nueva corónica y buen gobierno,* vol. 2, ed. J. Murra and R. Adorno [México: Siglo Veintiuno, 1980], p. 343.)

If a relationship can be made between one's media of communication and one's worldview, as many have postulated,[39] the Spanish could therefore be said to have had a visual, surface-oriented, and linear worldview, as opposed to the multidimensional and multisensory "worldview" of the Incas.

What made writing so radically novel for the Incas and their subjects, however, was its disembodied nature. Unlike writing, the quipu did not serve as a substitute for speech, as the data contained in it still had to be orally transmitted from one person to another. With writing, knowledge was no longer contained within human bodies but existed separately from them.

Soon after the arrival of the Spanish in the Inca empire tales spread among the Andeans of how the strangers "spoke in some white cloths, as one person would speak to another."[40] In this description of writing we find a suggestion of the isolation and alienation produced by literacy: People no longer speak directly to one another, but rather, "speak in white cloths."

This speaking in white cloths was so extraordinary for the Andeans that it was at first thought to be a supernatural endowment of the Spanish related to their purported abilities to eat gold and go without sleep.[41] One of the reasons the Andeans thought of the Spanish as "Viracochas" was because they were able to "name some of us by our names without anyone having told them to them, simply by looking at a cloth in front of them."[42] The Spanish were thus seen to be endowed with Viracocha's sacred naming power.

According to tradition, Inca Atahualpa threw down the book the Spanish priest offered him in their initial encounter after he had held it up to his ear and found that it said nothing to him.[43] While held captive by the Spanish, the Inca is said to have desired to penetrate the enigma of writing.

> Seeing the Spaniards read and write, he thought these abilities were innate among them. In order to verify this, he asked one of his Spanish guards to write the name of his God on Atahualpa's thumbnail. The soldier did so. When another Spaniard came in Atahualpa asked him, "What does it say here?" The Spaniard told him, and three or four others gave the same reply. Soon afterward, Don Francisco Pizarro came in . . . and Atahualpa asked him what the letters meant. Don Francisco Pizarro could not tell him because he did not know how to read. The Inca then realized that reading was not a natural ability, but an acquired accomplishment.[44]

Significantly, Atahualpa had the writing done on his thumb, as though trying to embody the disembodied word.

While the educated Inca elite may have recognized writing as an acquired accomplishment, and perhaps would even have made a voluntary transition from orality to literacy under more positive conditions, for the majority of Andeans writing remained a sorcerous power of the Spanish conquerors. A story was developed, in fact, that the Andeans themselves had once had writing, but that the Inca had forbidden its use after Viracocha said that writing caused plagues and great damage.[45]

The Andeans evidently associated writing with the devastation wreaked on their world by the Spanish. Atahualpa's inability to understand writing when first confronted with it, indeed, appeared to be the cause for the overthrow of the empire, for the Conquest began when Atahualpa rejected the book the Spanish priest gave him.

Writing did indeed play an important part in the Spanish campaign to incorporate the Andeans into the Spanish order. A seventeenth-century description of how visitors (inspectors working for the church) were to examine Andeans on their idolatrous beliefs, states, for example, that

> when the mass has been concluded, the visitor will install himself in the church with a table before him and a crucifix or cross on top of it. He will be provided with a blank book for the occasion . . . giving it this title: "Accusations of the Indians of Such and Such a Town, on such and such a day, month, year." And having a tax list of the town before him, he will call each one by name. . . . He will question each person briefly: Have you worshipped a huaca? And he will write down what he says. Have you had any dealings with a sorceror? Have you confessed to him?[46]

"Everything that is found out, by whatever means," says the author, "should be written down clearly," stipulating that "if there are any Indians who are literate, each one should bring what he knows in writing."[47] This, of course, was the prelude to the purging of all indigenous beliefs. One can well imagine the trauma that must have been produced in the native consciousness by this use of writing to abolish what the Andeans held sacred.

Lévi-Strauss states that "the primary function of writing is to facilitate the enslavement of human beings," and that "the struggle against illiteracy is indistinguishable, at times, from the increased powers exerted over the individual by the central authority."[48] The Spanish made full use of this function of writing in their domination of the Andeans. At the same time, however, the Spanish were aware that writing can also be used by the oppressed to develop and communicate an alternate order, and had mixed feelings about investing the Andeans with the potentially liberating power of literacy. Thus in early seventeenth-century Peru, when one Spanish priest was asked why he had not established a school for the native peoples under his care, he replied that "Indians did not need to read or write, since such knowledge would be no use to them except to criticize their priests."[49]

ANDEAN REACTIONS

One Andean who did use his knowledge of writing to criticize the Spanish was Felipe Guaman Poma de Ayala,[50] whose monumental work has furnished so much material for this book. The only way to prevent the Spanish from corrupting the Andeans, in Guaman Poma's view, was to keep them completely separate from the native population. In fact, he created a new metaphorical imperial body, with the Holy Roman emperor as the head, the "Christian king" on the right hand, and the "king of the Indies" on the left.[51]

Guaman Poma wrote his work in the hope that the Holy Roman emperor would read it and be so moved by it as to implement his ideas. In fact, his written pleas were no more attended to by the authorities than the vocal pleas of his compatriots. What is noteworthy for us, however, is that Guaman Poma adopted the Western view that writing expresses a more authoritative truth than speech. The following passage from Guaman Poma's work demonstrates his understanding of the supremacy of the written over the oral (and of the eye over the ear) in Spanish culture, and the untrustworthiness of the word of the Spanish.[52] "Justice must never be done by word, but by letter. . . . And if it is by word, don't listen to it and ask for it to be written. And in this way his Majesty will *see* and will provide."[53]

Nonetheless, Guaman Poma's work, although written, is heavily influenced by the structures of orality. He presents his arguments in defense of the Andeans in the form of an imagined dialogue between the Holy Roman emperor and himself, and even goes so far as to draw a picture of himself personally presenting his book to the emperor (fig. 21). Guaman Poma still conceived of knowledge as communicated through immediate active involvement, rather than through a process of detached observation and reflection.

The conflict between orality and literacy in the Andes was also a conflict between religions, that of the Andeans being centered on

Figure 21. Guaman Poma Presenting His Book to the Holy Roman Emperor (From Felipe Guaman Poma de Ayala, *El primer nueva corónica y buen gobierno*, vol. 3, ed. J. Murra and R. Adorno [México: Siglo Veintiuno, 1980], p. 897.)

oral dialogue with the sacred and that of the Spanish having its center in a sacred text, the Bible. Guaman Poma, a devout Christian, believed that literacy and Christianity were so intrinsically linked that the former was an essential requirement for the latter. "In this kingdom Indians, men and women, boys and girls, should know the language of Castilla, to read and write like Spanish men and women," for "whoever does not know how to [read and write] cannot be a Christian."[54]

Those Andeans who resisted such indoctrination defiantly proclaimed that "there [was] as much reason to believe their own ancestors and quipus and traditions as there [was] to believe the ancestors and writings of the Christians."[55] Manco, leader of the rebel Incas, urged his followers to reject the "false religion" of the Spanish and return to their huacas, saying "[the Christian God] is only a painted cloth. . . . The huacas speak to us."[56] In this proclamation we find a rejection not only of the religion of the Spanish, but also of their visual, surface-oriented culture.

The Spanish Conquest, nonetheless, profoundly disrupted the oral communication that animated the Andean cosmos. While Atahualpa was in prison he is reported to have sent his (Inca) priests to consult with the foremost huacas of the land as to the future of the empire. Despite all of the priests' entreaties, however, the oracles remained "deaf and dumb".[57] At first this was apparently thought to be a temporary and local condition, but then it was discovered to be general. Garcilaso de la Vega writes that "the [huacas] lost their power of public speech and could speak only in secret with great wizards, and even then hardly at all. . . . This occasioned a universal fear and astonishment among the Indians, who did not know the reason for the silence of their oracles, although they did not fail to suspect that it was due to the arrival of the newcomers in their land. They consequently feared and respected the Spanish more day by day as a people so powerful that they left their oracles without the power of speech."[58]

After the Conquest, therefore, the Andeans experienced a certain deafness and dumbness in their own cosmological order, while, according to the Spanish order, their idolatry and lack of writing made them blind.[59] Some Andeans experienced the sensory confusion that accompanied the clash between the two orders physically in their own persons. The turn-of-the-seventeenth-century Huarochiri manuscript tells the story of an Andean Christian convert, Cristóbal Choque Cassa, whose local huaca attempted to reclaim his allegiance through an attack on his senses. The *huaca* repeatedly blinded Choque Cassa with bright light and made his ears buzz. The Andean, however, was able to keep it at bay by proclaiming his faith in Christianity.[60]

Cieza de León describes a similar case occurring in 1549 in which an Andean chief, Tamaracunga, was tormented by "demons" who tried to prevent his conversion to Christianity.[61] These demons whistled and shouted at Tamaracunga and attempted to throw him off cliffs. The chief covered his face and body with blankets so as to hide from the demons; however, they still managed to fill his mouth with mud. The demons told Tamaracunga that they did this to stop him from becoming a Christian.

As Tamaracunga entered the church in order to be baptized, the demons appeared to him standing upside down, with their heads down and their feet in the air. Then "the demons . . . took him and held him in the air the way they were, head down and feet up. The Christians, saying loudly, 'Jesus Christ, Jesus Christ, be with us,' and crossing themselves, grabbed the Indian and put a stole [a vestment worn by priests] on him and sprinkled him with holy water. Still howls and whistles were heard inside the church . . . and [the demons] went to [Tamaracunga] and threw away the hat he had been covering his eyes with so as not to see them, and spit putrid, stinking saliva in his face."[62] When the mass was held the demons were finally silenced and Tamaracunga was able to be baptized.

A similar inversion of order was apparently sometimes used by the Incas to mark someone as an outcast. Guaman Poma remarks, for instance, that the soldiers of the victorious Atahualpa gave his defeated half brother Huascar rotten potatoes for food, urine for chicha, bitter weeds for coca leaves, a stone dressed like a woman for a woman, and so forth.[63] This was probably the sensory equivalent of being hung upside down.

In the case of Tamaracunga the "demonic" conquered Andean order is presented as inverted in contrast with the "holy" conquering Christian order. The upside-down position of the demons is contrasted with the right-side-up position of the Christians, the howls and whistles of the former with the prayers of the latter, and the demons' putrid saliva with holy water. The equilibrium of the old world had become disequilibrium in the new, and the sensory order of the old world had become sensory disorder.

Perhaps in order to prevent any reversion to the indigenous sensory and cosmological order, Guaman Poma proposed an ideal

Christian sensory order in the model prayer he wrote for his fellow Andeans.

> May Jesus Christ be in my eyes and in my sight. Cross.
> May Jesus Christ be in my ears and in my hearing. Cross.
> May Jesus Christ be in my nose and in my olfaction. Cross.
> May Jesus Christ be in my mouth, in my tongue, in all my words, in what I eat and in what I drink. Cross.
> May Jesus Christ be in my heart, in my soul, in all my thoughts and in all my deeds. Cross.[64]

Those Andeans who converted to Christianity soon discovered, however, that this was not enough to assimilate them into the Spanish order, for there was what appeared at times an almost complete disjunction between Christian ideals and Spanish practice. Guaman Poma sadly concludes that "the Indians of old were much more Christian. Even though they were infidels, they kept God's commandments and undertook good works of charity. Aside from their idolatries, they were Christian. In these times [the commandments] are not kept, although they should be kept more, since reading and writing are known and the Gospels and the laws of God are taught."[65]

Many Andeans rejected Christianity and the Spanish order outright. In the 1560s a millenarian movement arose among the native peoples of the Andes, beginning in central Peru, quite possibly in the rebel Inca stronghold of Vilcabamba, and extending to Lima and La Paz. Andean preachers proclaimed that, when the Spanish arrived, the Christian god had conquered the huacas, but that now the world was going to turn around and the Christian god would be defeated and the sea would rise up and drown the Spanish and flood their cities.[66] What these preachers were predicting was, of course, another pachacuti, and specifically, an uno pachacuti, world reversal by water, such as that which had marked the end of the first age of darkness and chaos in the Andean cosmogony.

The huacas were said to be hungry from the lack of ritual offerings, and to have left their rocks and other dwelling places and entered into human bodies—an example of the fluidity between land and body characteristic of the period of the pachacuti. Those persons who were possessed by huacas not only spoke for them, but also danced and had fits, for which reason the movement was known as Taqui Oncoy (Dancing Sickness). Molina of Cuzco writes that "there were many Indians who trembled and rolled on the ground, others threw stones like people possessed by the Devil, and made faces. When they were resting, people came up to them and asked them what they had and what they felt, and they responded that such and such a huaca had entered into their bodies. . . . The whole town would hold a fiesta for two or three days, dancing and drinking, invoking the huaca the person was said to have in his body and staying up all night."[67]

The continuous dancing that characterized the Taqui Oncoy movement was an important element in Inca rituals designed to ward off omens of death. Guaman Poma writes that, when faced with such an omen, the Incas danced all night "so as not to die."[68] The followers of Taqui Oncoy danced in order to keep the huacas and their old way of life alive.

Rumors spread among the Andeans that the Spanish needed the fat of Indians to cure a certain disease with which they were afflicted. Thus it was no longer only gold and silver the Spanish required to satisfy their aberrant physical needs, it was the bodies of the Andeans themselves (reinforcing the Andean association of the Spaniards with wild beasts). The huacas, however, were reportedly going to put worms in the hearts of the Spanish and their animals and Andean followers, which would cause them all to decay.[69]

In order to save themselves from the coming destruction, the Andeans had to renounce all things Spanish, including Spanish religion, Spanish clothes, Spanish food, and Spanish names, and return to the ways of their indigenous religion. Those Andeans who did not return to their ancestral beliefs would die and be forced to wander upside down with their feet in the air, or else be turned into animals.[70] In other words, those Andeans who refused to help set the world right again would be condemned to an inverted existence. That Christianity was influencing Andean religion even as it was being rejected, however, is revealed by the fact that the three leaders of the Taqui Oncoy movement took the Christian names of Juan, Santa María, and María Magdalena.[71]

The Taqui Oncoy movement was harshly suppressed by the Spanish, and the capture of the Inca holdout of Vilcabamba and the execution of the Inca Tupac Amaru in 1572 dealt it a final blow. As Molina of Cuzco, who was one of the priests accompanying Tupac Amaru at his execution, put it, "Our Lord, in his compassion, chose to enlighten those miserable souls; and those who are left have realized the absurdity of what had been preached to them and what they believed, on seeing the Inca dead and Vilcabamba [full] of Christians, and nothing having happened of what they thought, but rather the contrary."[72]

Christianity continued to be resisted by many Andeans after this time; however, this resistance no longer took the form of an organized religious movement. The execution of the last Inca decisively marked the division between the old, pre-Conquest way of life and the new, of which the Spanish and Christianity were an inevitable part.

The city of Cuzco was filled to overflowing with Spaniards and natives on the occasion of the execution of Tupac Amaru. Garcilaso de la Vega writes that

> when the Indians saw their Inca so near death they raised up such a cry of grief and pain that nothing else could be heard. The priests who were speaking with the Inca asked him to order the Indians to be silent. The Inca raised his right arm with his hand open, brought it up to his ear, then slowly lowered it to his right thigh. The Indians, understanding that they were being bade to be silent, ceased their shouting and crying and became so quiet that it seemed as if there were not a living soul in the whole city.[73]

At a sign from its conquered Inca, the Andean world was silenced to make way for the word of the Spanish world.

After the execution the Inca's head was impaled on a staff in the plaza. A Spaniard recalls that "each day it grew more beautiful. . . . At night the Indians would come to worship the head of their Inca, until one morning a certain Juan Sierra Sirujano looked out his window and saw the idolatry the Indians were committing. He passed on the information to [the viceroy] Francisco de Toledo, who ordered that head be taken down and buried with the body. . . . And so this undesirable adoration by the Indians of [their Inca's] head ceased."[74]

The breakup of the Inca body, in its actual, metaphorical, and sacred aspects, appeared complete. Nonetheless, despite the death of the last Inca and of the Inca state, and despite the disorder created in the indigenous culture by the Conquest and its effects, the basic cosmological concepts that were expressed by the Incas through the symbolic structuring of the human body would continue to survive in the Andes to the present day.

9
CONCLUSION

INCA COSMOLOGY AND THE HUMAN BODY

In Inca cosmology the structures and processes that governed the human body were also the basic organizing principles of the cosmos. The body was such a compelling and primary symbol for the Incas, in fact, that they applied it not only to the cosmos as a whole, but also to entities as diverse—and as apparently unlike bodies—as lightning, stones, and towns.

For the Incas, the Earth was a body that gave birth to the people, animals, and plants that lived on her. The Sun was a body that fathered the Incas, and that the Incas, in the words of Garcilaso de la Vega, "treated as corporeally as though he were a man like them."[1] The huacas were bodies that not only were given food and drink by the Incas, but also whipped when their communities of origin rebelled. When the Incas observed the stars, it was not only to revere them or to structure space and time, but also to establish a body-to-body, subject-to-subject relationship with them. The Incas, indeed, related to the cosmos through an overlapping series of bodies beginning with the body of the individual.

The system of reciprocity that was fundamental to Inca cosmology was expressed on a sensory level as an interchange of perceptions. In the Inca prayer cited by Pachacuti Yamqui, for example, the supplicant says to Viracocha, "by seeing you . . . you will see me."

In the Inca sensory order sight held first place, followed by hearing. These two senses were paired in a complementary duality, with the former being associated predominantly with the symbolic qualities of clarity and structure, and the latter with those of obscurity and fluidity. The complementary duality of sight and hearing was expressed, for example, in the Incas' use of only one term, illapa, to refer to both lightning and thunder. Garcilaso de la Vega writes that "when [the Incas] said, 'did you see the illapa?' they meant lightning; when they said, 'did you hear the illapa?' they meant thunder."[2]

Sight and hearing were apparently followed by smell, taste, and touch, in that order.[3] The primary function of odors in Inca ritual seems to have been to communicate the essence, or intrinsic quality of a thing; for instance, to convey the essence of a burnt sacrifice to the deities. The primary function of taste was to internalize the ideals associated with different flavors and foods. Touch, in contrast, was used to imprint structures on bodies and to dissolve the barriers between bodies.

It seems quite possible that the Incas, with their emphasis on spatial hierarchy, would have been influenced in this ranking of the senses by the position of the sensory organs on the body: the eyes at the top, followed by the ears, then the nose and the mouth. Touch might have been associated either with the hands or with the whole body.

The human face itself may have been thought of as a miniature model of the cosmos.[4] In this case the eyes (sight) would have been associated with the luminaries of the structured Upper World (the Incas, in fact, likened stars to eyes and pupils to birds),[5] the nose (smell) with the axis mundi of the mediating This (middle) World, and the mouth (taste, speech) with the opening to the fluid Lower World. If the Incas understood odors as traveling up to the brain, and flavors as traveling down to the stomach, this could have served in part as a model for ritually offering odors to the Upper World, and foods and drink to the Lower World. The ears (hearing), situated on the sides of the head, would be at once mediators between upper and lower and between front and back. The structure of the senses, therefore, can be seen as related both to the structure of the body and that of the cosmos in Inca cosmology.

With the arrival of the Spanish, the Incas were confronted with a radically different image of the cosmos and the human body. Those elements of Christianity the Spanish missionaries found the hardest to convince the Andeans to believe were those which were most disruptive of the indigenous structures that integrated human bodies into the cosmos; that is, that humans were all descended from one couple rather than from the earth of their homeland, and that the bodies of the dead were completely inanimate and had no role to play in the world of the living.

In the process of imposing their own order on the Andeans, the Spanish caused a mass disruption and destruction of Andean bodies: the body of the cosmos; the body of the Inca; the social body; the bodies of huacas; and the bodies of individual Andeans. The totality of the disorder produced by the Spanish Conquest was experienced by the Andeans as an inversion of the cosmos, a pachacuti, in which not only space was reversed, but also time.

With the reversal of position of the cosmic body, the future—the arrival of the Spanish—which had previously lain behind the body and been heard of only through oracles, became the present and the past (the Conquest period was associated with the first age), which lay visibly in front of the body. The Inca present and past, which had once been seen in front of the body, in contrast, soon after the Conquest became a time that was only heard of and that might return in the future behind the body.

The Andeans who participated in the Taqui Oncoy movement hoped to reverse the cosmos once again through another pachacuti that would exterminate the Spanish. The flooding of the Spanish and their cities that the Andeans believed would take place at this time, in fact, can be thought of as a ritual cleansing of the cosmic body, just as the Inca social body was ritually cleansed by water during citua, and the body of the individual after confession.

The Taqui Oncoy movement ended not with the righting of the cosmic body but with the breakup of the body of the last Inca. Nonetheless, the fact that the very disorder that was destroying Andean structures could be expressed by the Andeans through traditional symbols, such as that of the inverted body, demonstrates that Inca and Andean cosmology had the resilience to withstand and encompass to some extent even the devastation caused by the Spanish Conquest and the destruction of the Inca empire.

One of the great strengths of Inca cosmology was undoubtedly the coherence and comprehensiveness of its principles, which were applicable on both microcosmic and macrocosmic scales. Thus the union of a man and a woman in marriage, for example, was duplicated on a large scale by the union of the male sky and the female Earth, and governed by similar fundamental structures and processes. It is the logical consistency of Inca cosmology that has made it possible not only to examine the microcosmic role of the human body in Inca myth and ritual but also to extrapolate from that and other relevant data the macrocosmic significance of the human body in Inca thought.

The correspondence between the structures and processes of the body and those of the cosmos in Inca cosmology made the universal comprehensible by relating it to corporeal experience, and corporeal experience comprehensible by relating it to the universal. The Inca order itself was both externalized, incorporated into the cosmos; and internalized, incorporated into individuals, by the Incas through the dynamic corporeal and sensory dimensions of ritual.

All orders, nonetheless, from that of the cosmos to that of the body, were recognized by Inca cosmology as inevitably subject to disintegration. The Incas' attempt to preserve the bodies of their ancestors from the dissolution that naturally befell human bodies can be seen as symbolic of their desire to prevent their empire from succumbing to the dissolution that naturally befell social and cosmic bodies. It cannot have been unforeseen by the Incas, however, that even their world would eventually come to an end and be replaced by another, in accordance with the universal cycles of their own cosmology.[6] Thus, in a paradoxical fashion, the destruction of the Inca empire can be seen as an affirmation of the basic structures of Inca cosmology.

THE ANDEAN BODY TODAY

Throughout this work we have looked at some of the ways in which Inca beliefs and practices concerning the body continue to be evidenced in the traditional culture of the Andes of today. Let us now examine some of the ways in which present-day Andean attitudes to the body and the cosmos manifest the effects of the Spanish Conquest.

The Conquest itself is relived annually in various locations in the Andes in "Conquest plays." The dominant theme in these dramas is that of the disjunction between the Andean and Spanish cultures, and, in particular, between orality and literacy.[7] In one version the Spanish only move their lips when they speak, making no sound. They hand Inca Atahualpa a written letter, which he raises to his ear, trying to listen to its contents. When he is unable to hear anything he passes the letter around among his followers. No one can understand the mute message on this "maize leaf." One proclaims, "Seen from this side, it is like a swarm of ants. I look at it another way and I seem to see the tracks that birds leave on the muddy banks of the river. Looking at it again, I see stags, upside down and their feet in the air. And if one looks at it this way round, it is like a herd of llamas lowering their heads, with antlers like stags. Who on earth could understand that? No, no, my lord, I cannot possibly guess at its meaning."[8] When the Spanish priest hands Atahualpa a Bible the process is repeated. Atahualpa, unable to comprehend the new medium of communication, is beheaded by the Spanish and the world is shattered.[9]

The modern Andean world is still primarily an oral one. However, unlike the pre-Conquest era, when all of the cosmos was mediated by orality, modern Andean society is now marginalized within a dominant literate culture. For this reason many Andeans consider their world to be silent and still in comparison to that of the Incas. In one Peruvian community this was expressed as follows: "The Incas could talk directly to the Earth and Mountains and . . . herd gigantic boulders into fortresses. . . . Now that the Incas are gone, people have to move the rocks themselves and can communicate with the Earth and the Mountains only indirectly, through the medium of coca."[10]

A member of another Peruvian community comments that "the people of old used to help Ausangate [a mountain deity] quite a bit. They respected him very much. In those days they didn't go to mass and they weren't baptized. They simply spoke with Ausangate."[11] The transition from orality to literacy, from integration into the cosmic order to isolation, and from Inca religion to Catholicism, is movingly described in an account given by another contemporary Peruvian Andean. In traditional Andean fashion, the forces interacting in this myth are embodied and personified.

> God had two sons: the Inca and Jesus Christ. The Inca said to us, "Speak," and we learned to speak. From that time on we teach our children to speak. . . . The Inca built a tunnel in Cuzco through which he visited our Mother Earth. He conversed with her and took her gifts and asked her for favors for us. The Inca married Mother Earth. He had two children. . . .
>
> When they were born it made Jesus Christ very angry and unhappy. . . . The Moon took pity on him. "I can help you," she said and sent him a paper with writing. Jesus thought, "This will certainly frighten the Inca." He showed him the paper in a dark field. The Inca was frightened because he didn't understand the writing. . . . He ran far away. . . . He slowly died of hunger.
>
> When the Inca could no longer do anything Jesus Christ struck Mother Earth and slit her throat. Then he had churches built. . . .
>
> Ñaupa Machu [a survivor from the first age of savagery] was very happy to learn that the Inca was dead, for while he was alive he had had to remain hidden.
>
> Ñaupa Machu lived in a mountain called School. He was delighting in the blows Mother Earth had received when the two sons of the Inca passed by, looking for their father and mother. Ñaupa Machu said to them, "Come, I'm going to tell you where the Inca and Mother Earth are." The children happily went to school. Ñaupa Machu wanted to eat them. He said, "Mother Earth no longer loves the Inca. The Inca has become friends with Jesus Christ and now they live together like two little brothers. Look at the writing, it says so here." The children were very afraid and ran away.

> Since then all children have to go to school, and like the sons of Mother Earth, almost all of them don't like it and run away.
>
> Where are the two sons of the Inca? They say that when the elder is grown up he will return. . . . They say it's the children who have to find him. . . . But if we don't find him, he might die of hunger like his father. Will he die of hunger?[12]

To interpret: the Inca and Jesus Christ are able to interact with each other in this myth because they embody archetypes, the Inca that of Andean sacred structure, and Jesus Christ that of Western sacred structure. Both are seen as sons of God, mediators of the Divine, but in mutually exclusive ways.

The Inca animates and integrates the cosmos through speech: He speaks with the people and with the Earth. The relationship between the Inca—Andean sacred structure—and the Earth is one of reciprocity: He brings her presents and she grants him favors. Their marriage is a sign of the permanence and stability of this union, and their two sons proof of its dynamic power and cohesion. These two sons represent Andean culture, with its integration of the upper and lower halves of the cosmos.

Jesus Christ is upset at the birth of the Inca's sons because he realizes that there is no place for him in this structure. The Moon, symbol of sorcery, provides him with the means to silence the oral intercourse that binds and vitalizes the union of the Inca with the Earth: writing. The Inca flees the writing because it represents a way of life that is incomprehensible to him. He dies of hunger—the lack of ritual offerings and worship.

The cosmos is then plunged into a state of disorder in which Western sacred structure can devitalize the Earth and impose itself on her, and Ñaupa Machu, the personification of the chaos previously kept at bay by Andean sacred structure, is free to act. Ñaupa Machu resides within the school, which thus becomes the abode of chaos. He lures the two children—Andean culture—into the school with the promise of informing them of their parents, their past.

Once inside, however, the children find the forces of chaos overwhelming. Ñaupa Machu wants to eat them, to absorb them into himself. He tells the offspring of the Inca and the Earth that there is no longer any possible conjunction between the physical world and Andean sacred structure, and that the latter has been reconciled with Western sacred structure.[13] This is ratified by writing. Andean culture (the Inca's children), however, rejects the worldview put forth by anticulture and writing, and rejects the educational system of the West that promotes it.

For this reason, according to the myth, the children of the Andes dislike school, they sense that it tries to indoctrinate them with a hostile worldview, and they escape from it. The myth ends with the provision that it is only if Andean culture grows strong and is sought out by the children, that it can survive. Andean culture has been silenced but not destroyed. It remains latent, awaiting the time of its return.

This myth, like the Conquest plays, presents the post-Conquest world as one of bodily and sensory disorder for the Andeans. The integration of bodies through sound produced by Inca religion is contrasted with the disjunction of bodies through sight produced by Christianity. The myth declares that it is impossible for Christianity to exist in harmony with Andean cosmology (at one point the narrator says, "We have broken up with Mother Earth. Now Jesus protects us").[14] For Jesus Christ to triumph, the Inca and the Earth must die.

In fact, however, Christian and Andean sacred bodies and structures have inevitably become integrated to a certain degree through centuries of coexistence in the Andes. In the present-day Andes, the Virgin Mary, who is sometimes thought of as both the mother of Christ and the wife of God, is commonly associated with Pachamama, the Earth. God, in turn, tends to be associated with the sky, and particularly the sun. Christ is sometimes identified with God and the sky, and sometimes thought of as God on earth.[15] In this, Christ resembles the Inca who was associated both with the Upper World and with the middle, mediating world. Thus, although Christianity has introduced new elements into Andean thought, the cosmos continues to be ordered through the interrelationship of sacred bodies, as before the Conquest.

Humans are integrated into this order and into the social order through baptism. The Qollahuayas of Bolivia, for instance, think of baptism as a rite of passage from savagery to civilization, represented by the salt and baptismal dress used in the ceremony.[16] Unbaptized babies are believed by Andeans to cause hail which damages the crops.[17] Baptism, therefore, is necessary both for bodily and cosmic order.

The presence of Christian and other Western elements in modern Andean culture does not change the fact, however, that since the Conquest the Andeans have been marginalized from the dominant social order. The pachacuti that took place at that time, according to Andean cosmology, and displaced the Incas from the sky to the Earth, continues to be experienced by the Andeans. Many Andeans, particularly those of the Cuzco region, in fact, believe that the Incas now live under the earth or in the jungle.[18]

While the underworld and the jungle represent places of exile, they are also symbolic of the dark, fluid future. The Incas' presence there, consequently, indicates not only their exclusion from contemporary civilization, but also their potential for returning to power in the future. The expectation, indeed, exists among the Andeans that the Incas will return in another pachacuti, drive away the Spanish (most of whom are actually of mixed Spanish and Andean blood in the modern Andes), and establish an age of plenty and social harmony.[19]

In the tripartite division of time made by some Andeans this new age of the Incas is the last of three ages, known as God the Father, God the Son, and God the Spirit (Dios Yaya, Dios Churi, Dios Espíritu). The first age, of God the Father, is said to have been characterized by savagery. The second, present age, of God the Son, is characterized by civilization. In the third age, of God the Spirit, "we will all have wings and we will eat only fruit that grows on the bushes of the hills. Then the good will be sifted from the bad and only the good will remain, those who have followed what God has said. When the age of God the Spirit comes the Incas will come out of where they have been living in hiding in [the jungle]."[20] This expectation of the return of the Incas indicates that the Andean understanding of time is still cyclical to some extent, although it has certainly been modified by the influence of Christian concepts of history.[21]

The presence of the Incas as mythological figures in modern Andean culture is also found in legends concerning Inkarri, a culture hero whose name is composed of the words *Inca* and *rey* (king) and who orders the world as the Inca is remembered as doing.[22] David and Rosalind Gow have examined how the figure of Inkarri serves not only as an expression of modern Andean mythological thought, but also as an active symbol of social justice and a model for Andean leadership in contemporary Andean culture.[23]

Emiliano Huamantica, a twentieth-century Andean leader of Cuzco, for example, is remembered by Andeans as working toward the creation of a new Tahuantinsuyu "in which even the bitter black hearts of the mistis [mestizos] would be sweetened and we would all share a single, clean heart as in the time of our Inca forefathers."[24] In order to attain this new society, social solidarity is necessary: "He always said that we should walk as a single body, as a single man."[25]

Certain legends concerning Inkarri, however, describe the ruler's body as sundered, and powerless to act until it is reconstituted. Excerpts from three such legends are presented below:

> They say that only the head of Inkarri exists. It's growing downward, they say it's growing down to its feet. Then Inkarri will come back, when his body is complete. He hasn't come back yet.[26]

> They say he's in Cuzco now. We don't know who could have taken him to Cuzco. They say they took his head, only his head. And they say his hair is growing, his body is growing underneath. When it has reconstituted itself, perhaps it will be the Day of Judgment. . . . We don't know who killed him. Maybe the Spanish killed him and took his head to Cuzco. That's why the birds on the coast sing, "The King is in Cuzco." "Go to Cuzco," they're singing.[27]

> It was God (the Catholic one) who ordered the troops of the king-state to capture and decapitate Inkarri. . . . The head of Inkarri is in the Palace [of government] in Lima and it's still alive. But it has no power because it's separated from its body. . . . If the head of the god is freed and reintegrates with its body, it can once again confront the Catholic god and do battle with him. But if he is unable to reconstitute himself and recover his supernatural power, perhaps we will all die.[28]

Here the very life of the Andeans is presented as dependent on the reconstitution of the body of Inkarri. In keeping with the structures of Andean cosmology, Inkarri's body could be interpreted as symbolizing the body of the cosmos. The head that is kept captive by the Spanish, in this case, would represent the Upper World. The body, which must grow and be united with the head if cosmic order is to be restored, would represent the Earth, and the Andean peoples themselves.

EPILOGUE

What are the implications of Inca body symbolism for the fields of the anthropology of the body and the senses, and for religious studies? The Inca model of the body was, of course, a living and lived-in model, not a social-scientific construct. As a living model, however, it provides a basis for critiquing and enriching the theoretical models employed in Western social-scientific discourse.

The body symbolism of the Incas demonstrates the extent to which the human body is accessible to cultural mediation.[29] This cultural mediation involves not only body processes, but also body structures, along with body decoration. Douglas, in her anthropology of the body, has described the individual as possessing two bodies, one physical, the other social.[30] Among the Incas, however, each person could be said to have had two cultural bodies. One manifested the cosmic order in itself, that is, the complementary duality of left and right with their attendant values; the integration of parts through the circulation of fluids; the mediating role of in-between spaces; the disruption of structure caused by illness, and so on. The other located the individual within the social order: One's sex, for example, indicated one's social role; the design of one's clothes indicated one's community of origin; the quality of one's clothes indicated one's class.

The most important respect in which the Inca body model challenges the view of the body contained in Douglas's work is that the Inca cosmological order was not simply "inscribed" on the passive physical body (à la Douglas), it was *lived* and *thought* through the very entrails (sonco, or entrails and mind) of the body. In the body model of the Incas the cosmos, and to a certain extent society (humans, for instance, were created with their distinctive regional clothing already on) were built into the human body *at the time of Creation.* Thus the body was a presentation, not just a representation, of the cosmos.

Inca rituals sought to internalize the Inca social order in the bodies of individuals in the same way that the cosmic order was internal and intrinsic to the body. After all major rituals, for example, the participants were given sacred balls of yahuar sanco to eat, on the principle that these would physically order their bodies in accordance with Inca ideals. Once having eaten the yahuar sanco, an individual could not betray the Inca state without disrupting the integrity of his or her own body. At the same time, by actively engaging in the dynamics of Inca ritual, the participants externalized the Inca order. The bodies of the Incas and their subjects were thus not only inserted within the structure of the Inca state, they manifested that structure within themselves.

Michael Jackson criticizes Douglas for concentrating on the representational and ignoring the presentational aspects of body symbolism.[31] He is also critical of the nondialectical (or unidirectional) way in which she conceptualizes the relationship between the body and society, and he introduces the concept of "reciprocal anthropomorphism" by way of an alternative. According to Jackson, reciprocal anthropomorphism is a process by which individuals appropriate and internalize their physical and social habitus after already having experienced some form of identification with it. As Jackson puts it, "active realization of correlations between self and habitus presupposes an externalization which has already occurred, in which the self through bodily movement in concrete situations has discovered for itself an identity in relation to the immediate environment."[32]

The Incas' internalization and externalization of cosmic and social ideals through ritual resembles the reciprocal anthropomorphism described by Jackson. The participants in Inca ritual, however, did more than experience an identification with their habitus—society and cosmos—they enacted it within their own bodies. Indeed, for the Incas, bodily identification with and internalization of the cosmic and social order was but the preliminary basis that enabled humans to act as social and cosmic mediators.

It is impossible to adequately explore how the members of a culture "think through their bodies" without examining how they "think through their senses." Indeed, one cannot arrive at an understanding of how a culture perceives the world without first understanding how that culture perceives; for people do not perceive in an abstract way, but through their senses, and these are no less culturally mediated than any other part of the body.

Sensory perception formed an intrinsic part of Inca body symbolism, as well as constituting the very means by which it was possible to experience the body and the cosmos in the first place. The first dynamic experience humans have of their bodies in Inca cosmology occurs through their sense of hearing, when they are called into life by Viracocha. The distinctive gold ear ornaments of Inca men emphasized their privileged receptivity to the call of the divine. This superior receptivity of Inca hearing was complemented by the superior force of Inca vision: While ordinary Andeans dared not look directly at a sacred huaca, the Incas subjugated the huacas with their sight. The Inca order, therefore, was a sensory order as well as a bodily order.

The Inca body model is at once intriguingly like and unlike traditional Western concepts of the body. Two striking examples of this concern the relationship between the senses and the relationship between the sexes. In the West, for instance, we are familiar with the idea of sight being dominant and also with "sound-and-light" representations of the world, as in the cinema. What strikes Westerners as unfamiliar, however, is the Inca notion that a visual representation cannot be sufficient in itself, that it needs to be coupled with orality to be complete.

As far as a Westerner is concerned, a religious icon, for example, although a purely visual model, is complete. Any other sensory manifestation it might display—an odor, say—is an "extra." From the Inca "point of view," however, the religious images of the Spanish were mere painted cloths not worthy of reverence because they could not speak. To be unable to speak and to hear in Inca cosmology meant quite simply to be excluded from participating in the cosmic order. Thus, the Incas endowed all the elements—"bodies"—of their cosmos with the power of orality.

Again, in the West we are more than familiar with the notion of men being superior to women and of each sex having different proper spheres of action and different symbolic associations. In Western tradition, however, the male principle has been treated as the major principle within which the female principle was subsumed. A daughter, for example, was subsumed by her father, and a wife by her husband. Among the Incas, in contrast, male and female were considered to be too intrinsically different for one to be represented by the other, and thus constituted interdependent but separate orders.

In Inca cosmology there were parallel orders of male and female bodies on the cosmic level (e.g., the sun and the moon), on the social level (e.g., the priests and the acllas), and on the individual level (e.g., the Inca and the Coya, husband and wife). While male bodies were deemed superior to female bodies, each was essential for the constitution of the cosmos and society. The complementary duality of the two sides of the body made it possible for the Incas to conceive of the cosmos and society in terms of two parallel, complementary orders integrated within one totality.

The model of the body presented in Inca cosmology invites comparison with the body models implicit and explicit in the cosmologies of other cultures. The exploration of how meaning is encoded and expressed in the human body can provide important new insights into religions that have traditionally been studied through their texts. In particular, however, it can be a valuable means of studying the religions of oral peoples, whose sacred ideals are not embodied in texts but in human bodies.

The sensory symbolism employed by different cultures can provide an especially rich field of exploration.[33] Scholars concerned with the cultural and cognitive effects of literacy (e.g., Walter Ong and Jack Goody) have tended to group nonliterate cultures indiscriminately together as oral-aural cultures. The fact that a culture is not literate, however, does not necessarily mean that it will value hearing above sight. The Incas, while an oral culture, in fact gave a higher symbolic value to sight than hearing. Oral (and literate) cultures can thus display a wide variety of sensory models.[34]

Furthermore, one cannot assume, as Ong does, that certain cultural effects will follow from a particular sensory order. While each sensory modality may have specific inherent properties—for example, sight is unidirectional, hearing is multidirectional[35]—cultures remain free to selectively emphasize these properties and imbue them with their own values. Thus the visualism of the West, as we have seen, is quite different from the visualism of the Incas. Sensory symbolism, therefore, must be understood within the context of the particular culture where it occurs.

It is often in situations of culture contact that basic differences between cultural evaluations of the senses are revealed. We have looked at how this occurred in the Andes with the confrontation of Inca and European sensory models at the time of the Conquest. A similar confrontation of indigenous ways of sensing with European ways occurred in North America. Consider the following statement made in 1855 by an Indian chief involved in treaty negotiations with the government of the United States. "Us Indians are blind. . . . We do not see the earth well, the Lawyer sees clear. The reason that I do not know anything about the ground is that I do not see the offer you have made as yet. If I had the money in my hand then I would see. . . . I wonder if this ground has anything to say. . . . The Earth says that God tells me to take care of the Indians."[36] Observe how the chief characterizes the Indians as blind in contrast to the whites, and particularly to the literate lawyer who "sees clear." The chief has grasped, however, that in the socioeconomic structures of the white world the ability to see is dependent on the possession of money: If I had money, I would see. After acknowledging that he cannot see as whites do, the chief returns to his oral world, which is animated and structured by speech. The Indians "do not see the earth well," but they can hear it.

In closing, it should be said with regard to the body and sensory symbolism of the Incas that there are many intriguing and important questions still left unanswered. How did the body and sensory order of women differ from that of men? What alternatives to the dominant body model were presented by ethnic minorities? How was the bodily experience of those who did not fit the normative body model—dwarves, the blind, the deaf, and so on—ordered? The lack of information on subordinate groups in the Inca empire means that such questions can probably never be answered. In the modern Andes and in other cultures, however, these and other questions await exploration.

APPENDIX

QUECHUA BODY AND SENSORY TERMS

Following is a compilation of the Quechua body and sensory terms contained in Diego de González Holguín's Quechua-Spanish dictionary of 1608. This colonial Quechua body nomenclature is supplemented with a modern (Cuzco) Quechua vocabulary of body terms taken from *La rebelión de los mallquis* by Ángel Avedaño.[1]

The colonial Quechua terms are spelled according to González Holguín's (not always consistent) orthography. My English translations of these terms are from González Holguín's Spanish translations of the Quechua. Where of interest, a literal translation of the Quechua term is provided in brackets after the word.

While González Holguín's dictionary is the most extensive (some seven hundred pages) of the Quechua vocabularies produced in the colonial period, it is far from exhaustive, and many specialized Quechua terms are lacking. Enough body and sensory terms are provided, however, to make it possible to construct the basic body and sensory nomenclature of the Incas.

The first lists in this appendix are of colonial Quechua terms for external and internal body parts. Several points are worthy of note concerning this body nomenclature. While most of the body divisions listed here correspond to those with which we are familiar in the West, there is only one term, *maqui,* for both hand and lower arm, and another single term, *chaqui,* for both lower leg and foot. There are, however, separate terms for the upper arm and lower arm and upper leg and lower leg. This indicates that the division of the limbs into upper and lower halves was more important to the Incas than an arm/hand or leg/foot division.

As well, many parts of the body have metaphorical connotations. The correspondence of the body to the land in Quechua thought is seen in the application of body terms to features of the landscape. *Uma,* for instance, means both "head" and "mountain peak," while *ciqui* means both "buttocks" and "mountain slope." Certain parts of the body are identified with the functions they perform; thus the mouth, *simi,* is also "speech," and the throat, *cunka,* is also the "voice." Other parts of the body stand for aspects of the personality. *Tullu,* bone, is used to refer to a person's "physical" character in expressions such as *quella tullu* (lazy bones or idle person), while *sonco,* the heart and stomach, refers to a person's "spiritual" character, in terms such as *llullmi sonco* (deceitful heart or person).

The only terms for "soul" given by González Holguín are in Spanish: *alma* (soul), and *ánima* (spirit). This reflects the efforts of the church to suppress indigenous concepts of the "soul" in the Andes and replace them with Christian ones. Quechua spiritual terms that are given are *cay* (being and essence) and *callpa* (the strength, potency, and faculties of the body and spirit).

The second set of lists provides colonial Quechua terms for sensing, and colonial Quechua terms of sensory description, including those dealing with sensory acuity and loss. The senses are divided according to the classification given in González Holguín's dictionary, that is, into sight, hearing, smell, taste, touch, and intelligence. In making this classification, González Holguín may well have been influenced by the contemporary Western model of the senses, in which the faculties of the mind were characterized as "interior senses." Nonetheless, Quechua terms are provided for each of these six faculties and for none others (such as kinesthesia).

In Quechua a number of variations on a basic verb can be expressed through the addition of suffixes. For example, the suffix *payani* can be added to a verb stem to convey the meaning of doing the action in the verb excessively: *Kahuani* (to look) becomes *kahuapayani* (to stare). The subtlety of meaning this allows for will be noted in the variety of verbs based on a single root listed for the different senses.

With regard to the sensory terminology, the large number of descriptive terms listed for vision and color, and for smell is of particular interest. These sensory fields were evidently well elaborated by the Incas. By comparsion, there are relatively few descriptive terms for flavors and textures. Gustatory and tactile terms are used synesthetically, however, while other sensory terms are generally not. *Mizqui* (sweet), for example, is used to mean "pleasant" or "soft" when applied to odors, sounds, sights, and textures; and "pleasant" or "deceitful" when applied to abstract concepts, such as a person's character. Likewise *llampu* (soft) is used generally to mean "mild." Thus, for example, in Quechua *mizqui sonco* (sweet heart) is someone with an amiable disposition, while *llampu sonco* (soft heart) is someone with a gentle disposition. This indicates that touch and taste provided general terms of description for the Incas, while the other senses had more specific meanings.

Quechua auditory terms, aside from those dealing with speech, of which there are a lengthy number, seem often to be based on onomatopoeia. Doubtless there were many more such onomatopoeic terms in common use than those González Holguín included in his dictionary.[2]

No one sense is explicitly associated with intelligence in González Holguín's vocabulary. There are sensory connotations to certain terms dealing with intelligence; for example, *happini* (to grasp) can be used figuratively to mean "to grasp an idea"; *unanchani* (to understand) also means "to make signs"; while *upa* (deaf-mute) also means "to be stupid." No sense, however, appears to have been used as a direct metaphor for understanding or knowing. Reason thus constitutes a separate faculty in Quechua.

The *sonco* (heart, stomach, and entrails) is the primary location of the mind, as well as the emotions, in Quechua thought. The head *(uma),* however, is used as the locus of the memory. The Quechua term for memory, in fact, is *yuyana huma,* "head thought," while that for understanding is *sonco yuyana,* "heart thought." This is, no doubt, a simplification of what in Inca culture were complex notions of intellectual and bodily processes.

The list of modern Quechua body terms at the end of this appendix provides a comparison with the terms compiled by González Holguín. It will be seen that the modern Quechua terms are very similar to the colonial ones. (A notable exception is the modern Quechua word *kurku* (body), which is taken from the Spanish for body, *cuerpo.)* This consistency of terminology, in spite of centuries of Spanish influence, demonstrates the importance of indigenous body nomenclature to Andean thought.

Colonial Quechua Body and Sensory Terminology

The Body

External Body Parts

ucu	body
ucuycuna	members
aya	cadaver
paña	right side
lloque	left side
ñaupa	front
quepa	back
ccara	skin

Head

uma	head (also mountain peak)
umap pucyun [head fountain]	crown of head
chuccha	hair of head
capra	facial hair
matti	forehead
utina [stunning place] or *huañuna* [fainting, dying place]	temples
uya	face
ñaui	eye (also face, flat side, point)
ñauip pichiun [eye-bird]	pupil
ñauip yurakni [eye-white]	white of eyes
ñauip ccaran [eye-skin]	eyelid
qquesipra	eyebrow
pullurqui	eyelash
rinri	ear
llutu	earlobe
cencca	nose (also hill top)
cincap ñaupin [nose-eye]	tip of nose
caclla	cheek
simi	mouth (also speech)
virpa	upper lip
cirpi	lower lip
quiru	tooth
tokma quiru or *chocta quiru* [disproportionate tooth]	canine
maray quiru [mortar tooth]	molar
huacco roquiru [separate tooth]	wisdom tooth
kallo	tongue
kallup circan [tongue vein]	frenem
cunca	neck, throat, or beard (also voice)
toncor	throat

toncor muco [throat joint or knot]	Adam's apple
muchu	back of neck

Torso

cazco	chest
ñuñu	breast
huacta	flank (also side)
huassa	back
huassuman	shoulder blades
ccarmin	shoulder blade
pacca [hiding place]	lap
uzpun	belly
pupo	navel
chaca	hip (also bridge)
ciqui	buttocks (also bottom and mountain slope)
tiyerik pullpullun necta [hair of seat]	pubic hair
raca	vagina
ullu	penis

Limbs

mucu	joint
rikra	shoulder
huallhuacu	armpit
maquip ricran	arm including the shoulder
cuchuch [corner]	elbow
machin [condiments] or *mizquin* [sweet]	fleshy part of the arm or leg
maqui	hand and arm up to the elbow
chocmi	fist
panpa maqui [flat hand] or *tanta maqui* [gathering hand]	palm
rucana	finger
mama rucana [mother finger]	thumb
tocsik rucana [pointing finger]	index finger
chaupi rucana [middle finger]	middle finger
siui rucana [ring finger]	ring finger
sulca rucana [smallest finger]	little finger
mucu mucu	knuckles
sillu	nail
chaqui	leg and foot
chanca	thigh and hip or whole leg
chuppan or *pusta*	calf
chaquip mucun [foot joint]	top of the foot
pichhusqui chaqip mucun [bone of foot joint]	ankle

chaquip ñaupin [foot-eye]	tip of foot
panpa maqui [flat foot] or *tanta chaqui* [gathering foot]	sole

Internal Body Parts (skeletal)

tullu	bone
chillina	marrow
ayapuman [cadaver head]	skull
karmin	shoulder blade and/or humerus
huassa tullu [back bone] or *huassa mucco mucco* [back joints]	spine
huamaniy	bone of the stomach opening *(hueso de la boca del estómago)* or shoulder blade
huacta	rib
cazcohuactan [chest ribs]	front ribs
huassahuacta [back ribs]	back ribs
ppiruru	kneecap
pichu	bones of the hands or feet

Internal body parts (soft)

aycha	flesh
uccup hillin	body juice
aychap hillin [flesh juice] or *hoycocaynin* [wetness]	humors
vira	fat
circa	vein
circa tticticcniynin	pulse
mamancircca [mother vein]	artery
yahuar	blood
anccu	nerve
mucu	joint
ñutcco	brain
çancar	trachea
çanka	palate
ttoccay	saliva
unan callo	uvula
chaca	tonsils
millpuna	esophagus
macchin [seasoning] or *mizquin* [sweet]	biceps
ñuñu	milk (also breast)
machca or *kayan*	spleen
sonco	heart, stomach, and entrails (also mind)
vicça	stomach

chhunchulli	entrails and bowels
aca	excrement
cucupi	liver
hayaqquen [bitter]	gall bladder
huassa ruruy [back fruit]	kidneys
puccuchu [bubble] or *yspaypuru* [urine cup]	bladder
ispay	urine
huahua tiyana [baby seat]	uterus
sullu	fetus
ttamin or *yuyun*	placenta
yumay	semen (also beesting)

The Senses

Sight

ricuna	the sense of sight
ricuni	to see
ricurayani	to be on the lookout
ricuchacuni	to look everything over
ricuccuni	to be watching
ricuchini	to look a lot or too much
ricumuni	to go and see, visit
ricupunacumuni	to go and see each other
ricutamuni	to see in passing
ricumuni	to see someone who has been absent
ricumpuni	to return from seeing something
huañuy huañuyta ricuni [deadly sight]	to look angrily and deeply
kahuani	to look
kahuapayani	to stare
kahuakahuani	to look many times or constantly
kahuarccuni	to look inside or outside or up, to look over
kahuaycuni	to look down or inside
kahuarini	to look up or to recover one's sight
kahuapayani or *ricupayani*	to watch over something, to guard
kahuapuni	to look for something lost
kahuaycucuni	to look at oneself
kahuachacuni	for a group of people to look at something, or to look at various things
kahuapayani or *ricupayani*	to look for someone to help
catirayani	to watch where someone goes
caructarac kahuarini	to look into the distance
muzpanaycama kahuani	

[to lose one's senses looking]	to look until one's sight fails

Hearing

uyaricuna	the sense of hearing
uyarini	to hear
uyaricumun	to hear at a distance
uyaricuchani and *uyarirayani*	to be listening
uyaricumuni	to go and hear something, to find out
uyaripayapuni	to hear a secret or to hear too much
mizqui mizquilla uyarini	to hear with pleasure
uyaricupumuni	to hear in passing what others are saying in secret
uyaricupuni	to secretly listen
uyariusipucuni	to listen to what everyone is saying and make interjections
uyarichini	to make someone else hear something, to tell

Smell

mutquicuna	the sense of smell
mutquini	to smell something
mukacuni	to smell a good odor
mutqquichini	to make someone else smell something
mutqquichacuni	for a group to smell something together
mucacumuni or *mutqquimuni*	to secretly sniff out what is being planned
çamaycuni	to come across the vapor or odor of food or drink, to inhale (also to inspire)

Taste

mallicuna	the sense of taste
mallini	to taste
macchicani	to savor, to enjoy a food
machikllicani	to try a food
llakhuani	to lick, to try, to taste

Touch

cullanacuna or *chancanacuna*	the sense of touch
cullani	to feel pain or anything through touch
cullachini	to tickle

cullacuni	to be tickled
cullachinacuni	to tickle each other
cullachiytucuni	to be tickled against one's will
cullachicuni	to be tickled with one's consent
cullachicucuni	to tickle oneself
chancani	to touch gently
llamacani	to touch, to feel
llamcanacuni	to touch each other
llamcanacuni	to handle
llamcaycuni or *chancaycucuni*	to touch oneself gently
llapicuni	to press, to feel something with one's hand
llapini	to squeeze, to break something
llapipayani or *chancapayani*	to touch roughly, to handle
haptani or *happini*	to grasp
llullaycuni	to caress, to pat
llullapayani	to deceive or ask with caresses
makallini	to embrace (also, *muchani:* to kiss; *kapini:* to knead; *azpini:* to scratch, etc.)
cconini	to feel very hot
cconicuni or *macacuni*	to warm oneself
cconiyani	for one to become warm
cconichacuni	to warm oneself all over
cconurayani	to warm oneself for a long time
rupaypacuni	to warm oneself
rupacuni	to burn oneself, to be hot
macaycachani	to warm oneself often
macarayani	to be warming oneself up too much
chiriyacuni	to feel cold
chirayani	to always feel cold
rittirayani	to be cold as snow
chiriyani	to freeze
nanani	to feel pain or to feel sick
nanatamuni	to travel with pain or sickness
nanaycachani or *nananani*	to go around with pain or sickness
nanarcarini	to suffer many pains or sicknesses
muchuni	to suffer

Intelligence

yuyanacuna	the interior senses—reason, imagination, and memory
yuyani	to think, to imagine, to remember, to be responsible for something
yuyacuni	to be thinking something
yuyacucuni	to be thinking to oneself

yuyayuyani or *yuyarayani*	to constantly remember something
yuyarccuni or *yuyarimuni*	to recall something
yuyarcupuni	to remember something completely
humaypi apaicachani [in head to bring often]	to try and bring to mind
huaynanaricuni	to remember very much what one loves
hamuyniy [*hamu:* mode or type]	judgment and will
hamuni	to plan or order something
hamuycupuni	to plan something for the good of another
hamupayani	to conjecture, to be indeterminate
hamutani or *huatuni*	to guess correctly
hamurpayani	to understand, or consider deeply
hamutani	to reason, to organize
unanchani [*unancha:* sign]	to understand or consider, to make signs
riccharini	to come to one's senses or remember on waking up
ccutcuni [to gnaw]	to reason
yachani	to know
yachacuni	to learn
yachapacuni	to learn through imitation
yñini [*y:* it is true]	to believe
yñirini	to begin to believe
yñiusini	to believe with others or what others believe
yñinacuni	to believe each other
checcamnini [*checca:* singular, separate, straight, true]	to recognize as true
checcanchaccuni	to figure out correctly
achnini [*ach:* it is probably so]	to believe or assent with doubts
acharinini	to believe or assent with little doubt
achnirayani	to remain doubtful
manaynini or *manachnini*	not to believe or assent
cconccani	to forget
cconccayani	to slowly forget
cconcaycupuni	to forget about another to his harm
cconccapayani	to forget too often
muzcuni	to dream
muzcupucuni	to dream of another
muzcupayani	to dream too much

General visual terms

ricuy ricuylla	something to see, or with a pleasant appearance

ramca or *samca*	dream visions or phosphenes
ricurini	for something lost to appear
ricurimuni	for someone who has been absent to appear
ricurimun	to appear in the distance
ricuchini	to make someone see something, to give sight to the blind
huallparicuni	to adorn and dress oneself beautifully
mircatan [*mirca:* freckle]	to look good
huahuay çumak runa [childish beautiful person]	one who looks good but has no courage or wisdom
ccanchariy	light
ccancharini or *ccanchani*	to light up, to show off one's clothes
ccancharcarini	to emit rays
huachiynin [*huachi:* arrow]	ray of light that enters through the window
yntihuachin	ray of sun
ynti hinasutilla	as bright as the sun
ynti hinacumaklla	as beautiful as the sun
quillap quillariynin [*quilla:* moon]	moonlight without rays
quilluy	the light of the moon covered with clouds
quillani	for the stars or the clouded moon to give light
cuyurpa quillariynin [*cuyur:* star]	starlight
cituynin	firefly light or reflected light
lliukniynin	lightning light
yllarina [*ylla:* sacred relic; *yllapa:* lightning]	to shine, to flash, to light up
cittuni	to shine, give off light
cachachachanmi	for the stars to flash
llipiyanmi	to shine, to reflect light
cachachacha	spark
cachcachnini	to throw off sparks
llipiyani	to shine like new
llipipini	for smooth things to reflect light
chipipipini	to reflect light, to show off new clothes
pacsa	beautiful and shining
pacsani	for the moon to shine without rays
pacsaruni	for the moon to appear from behind clouds
curumpini	for light or snow to dazzle

ccananananak rupaypi	for heat to shimmer and dazzle
ciçamana [*ciça:* flower]	something that looks as beautiful as a flower
rakrayak [*rakra:* crack] or *tutayak* [*tuta:* night]	darkness
chhisiyan [*chhisi:* evening]	to grow dark in the evening
tutayan	to grow completely dark at night
llantu	shadow
llantunllantuyan	to make a shadow
llantuycupini	to shade one desirous of shade
llantuycupayani	to shade one not desirous of shade
llantuicucuni	to shade oneself
llantuyrayanpacha	to shade for a long time

Color terms

rikchay	color, appearance
llimpi	color, red
ppata	of a single color
ppatayuyac	all white; *ppataquella:* all yellow, etc.
paucar	bright hue; *paucar puca:* bright red; *paucar cani:* bright brown, etc.
paucarccuna	a variety of colors, e.g., of flowers or feathers
pauccar richik	colors in general
pauccar richac inquillcuna [brightly colored flowers]	flowers of all colors
pauccar richac patpacuna [brightly colored feathers]	feathers of all colors
nina [fire]	bright; *nina llimpi:* bright color; *nina quello:* bright yellow, etc.
llanca	very deep hue; *llanca puca:* very deep red; *llanca yana:* very deep black, etc.
cutisca:	dull, discolored; *cutisca paco:* dull bay, etc.
kamparmanani	for a color to be worn out
huc ricchaymanani [another color to appear]	to change color
mircca mircca [freckles]	variegated
muru muru [spots]	variegated or dappled with colors
pirascca [*pirani:* a sacrificial ceremony]	striped with colors or blood
missa	bicolored
allca or *ticlla*	white and black
pillaca	purple and black
yana	black
yanayachini	to blacken, dirty

yanayani	to become black
yanachanani	for something with a good color to turn black
colli	dark brown
çani	dark brown
mapatayna	dark reddish brown
chumpi	dark reddish brown
paco	bay
chhekchi	light brown
yurakcani	light brown
oqque	light brown or grey
çaniyan	to appear brown
puca	red
anquepuca	very red
yahuarpuca	bloodred
ychma puca	red like the ychma fruit
ayrampu	a red dye
macnu	a red dye
mantur	red color of a fruit
pucaymanan	to be very red
pucayan	to redden
pucayani	to become red
pucayachini	to make something red
panti	reddish orange
uila api	orange
quellu	yellow
sunchu quellu	very yellow like the sunchu flower
ychma quellu	yellow like the ychma fruit
quelluchanani	to become yellow
quelluyachini	to make something yellow
ccomir	green
ccomiryani	to become green
ancas	blue
ccopa or *yurak ancas*	light blue or turquoise
yurak	white
cuyllu	white
huañuycullu [deadly white]	extremely white
rittiymanani	snow white
llumpaclla [pure]	all white
yurakchani	to whiten
quespi	something transparent
chuya	clear

Auditory terms

uyaricumuni	to sound
uyarichini	to make hear, to tell

cunca	voice; *pissi cunca* [little voice]: quiet voice; *hatun cunca* [great voice]: loud voice, etc.
rimani	to speak
rimarini	to speak a little
rimacurini	to speak angrily
rimarcuni	to speak hastily and regret it later
rimapayani	to talk too much
rimarcarini	to speak nonsense
rimarcucarini	to criticize too much
rimacuni	to speak ill of another
rimaycucuni	to talk to oneself
rimachini	to discuss business
rimachichini	to persuade someone to tell something or to speak out against an evil so that it may be corrected
rimachicuni	to solicit women
rimachipuni	to speak for another
rimachacuni	to joke
rimachitamuni	to say something on departing
napaycunacuni	to speak in greeting
huatatatani	for many to speak together
aclluycachani	to stutter
simi ucupi rimani [mouth inside to speak]	to mutter
rokyani	to shout and make noise
royaycachani	to shout out often
rocyapayani	to disturb others by shouting
rocyachacuni	to shout wildly
caparini	to shout
huakyarini	to call out
kaparini	to cry and call out
achini	to sneeze
hahanini	to laugh
hiqquini	to hiccough
taqui	music and dance
taquini	to sing
huaccan	for animals to bark, bray, sing, etc.
huanhuannini	to bark
huaccarachani	to howl
huacani	to cry
huarpicuy or *uichchichicuy*	for birds to sing pleasantly
kacniy	to thunder
cunununun	to make a loud sound, to thunder
sallallallan	to make a crashing sound
ppakakakani	to pop, to bang

chhacacacan or *roccoccoccon*	to make a breaking sound
cakaccakan	to rattle, to rustle
razniy	the sound of bones
chiuiuinichini	for leaves to rustle in the wind
ppatpatnini	for birds to make noise with their wings
chusnin	to sizzle
huanhuanmi	to buzz
huactani	to drum
pirutuni	to play the flute
kuyuyni	to whistle
uichichichini	to chirp
qquiqquiqquiqquin	to creak
chhukninin ynin	silence
upallani [to be mute]	to be silent
chhukninchini	to make silent

Olfactory terms

mucani	to emit an odor
yachachicusca aznaycuna [created odors]	compound odors
mukay	fragrance
mukani	to emit a good odor (also to make holes in plants so that they absorb more water)
cuntuyaznan or *cuntun*	good-smelling things
alli aznak [good odor]	something that smells good
cuntuy aznay [good odor] or *mizqui aznay* [sweet odor]	fragrance
cuntuy aznarcun	to emit a good odor
cuntuni	to smell good
cuntutamun	to leave a good odor
allinta aznan [good smell] or *mizquilla aznan* [sweet smell]	to smell good
çamarccuni	to evaporate, for a good smell to lose its strength
mizquilla aznaycachan	to sometimes give a good odor or to go around smelling good
mutqquichini	to make someone smell something
aznachicun	to have oneself or let oneself be smelled
aznay	stench
huañuy aznay [deadly odor]	a terrible smell
aznani	to smell rotten, to smell bad
aznarinnam	to begin to smell bad
aznapayani	to give a bad odor to another

aznak, curayak, or *ccapak*	something that stinks
aznarayak	something that always stinks
aznay capa	something full of foul odor
ccuncu	something that smells naturally bad
ccuncucay	to smell naturally bad
ccuncuksimi [stinking mouth]	someone with foul breath
ccuncucapa huayllaco [stinking armpit]	someone with armpit odor
curayani	to smell bad
ccapani	for something burning to smell bad
ccapan	to give off a bad odor that makes one sneeze
lleqqueqqueqquen	for rotten meat to smell bad
ccapachini or *coznichini*	to cense
ttanta ttantallam aznani [bread smell]	to smell of bread, *ccorpa*
ccorpam azṅani [earth smell]	to smell of earth, etc.

Gustatory terms

mizqui	sweet, tasty
ñucñu	sweet
huañuy ñucñu [deadly sweet]	extremely sweet
cachi	salt
cachiyoc	salty
puchco	sour
hayac	bitter or peppery
huañuy hayak [deadly bitter]	extremely bitter
ccatque	unripe taste
ccayma	insipid
mizquin	spices (also the fleshy part of the arm or leg)
mizqquillicuni	to taste good (also to give pleasure)
mizqquichicuni	to eat something tasty (also to do something with pleasure)
mizqquichani or *ñucñuyani*	to sweeten, to flavor
machitmachini	to season (also to penetrate, physically or spiritually)
machitmacuk	for food to be taking on flavor
machitmascca or *mizquin chascca*	well-flavored food
ppoccossca	well-flavored drink

Tactile terms

catccu	rough
anakcay	hard
llampu	soft, smooth
llullu	tender
ñupulla	soft to eat
kaypa	something soft to eat or to work

catccuyani	for what is soft to become rough
rupay	heat
rupayninchic	natural heat
rupaynin	heat rays of the sun
rupapayan	for fire or the sun to burn someone
huañuy rupay [deadly heat]	excessive heat
timpuk	boiling hot
chiri	cold
chirinmi	to be cold
chirichini	to cool off
chiriyachini	to put to cool off

Intellectual terms

checca or *sullull*	true
llulla, nucna, or *pallco*	false
yachaypac usanalla [to know easy to do]	easy to know
ttitu	difficult to know

Sensory acuity

ccazccaruna [sticking person]	one who uses all his senses sharply and subtly
ccazccactam ricuni	to see subtly
ccazccactam uyarini	to hear subtly
ccazccactam mutquini	to smell subtly
ccazccactam mallini	to taste subtly
ccazccactam llancani	to touch subtly
ccazccactam yuyani	to think subtly
ccazccactam unanchani	to understand subtly
ancha cullak [very sensitive]	to be very sensitive to pain
ccaraktullu [chafed bones] or *cullaktullu* [sensitive bones]	one who is very sensitive to pain
chiricullak	one who is very sensitive to cold
ricuyta ancham usachicuni [great sight to acquire]	to have keen sight
soncoyok [one with a heart]	one who has the use of his reason
kallu runa [ingenious person]	an ingenious person
yachapa	a quick learner
sonco capa [great heart]	a wise person
amauta	a sage

Sensory loss

ayarayani	to be senseless like a cadaver
natarayani	to be injured and unable to move
chiranayani or *nucuyani*	to be paralyzed
uticayani	for a limb to fall asleep
huañuni	to die or to faint

machani	to become drunk
huiccaccuni nam [to diminish]	to lose one's strength and faculties with old age (applicable to each sense)
ñauca	blind
ñaucayani	to be losing one's sight and understanding
ñaucanayani	to be on the verge of being blind
ñaucayachini	to blind someone, physically or spiritually
ñaucachanani	for one who was not blind to become blind
hapra	shortsighted
haprayani	to become shortsighted
upa	mute, deaf because of old age (also stupid)
roktto	naturally deaf
rocto rocto tucuni	to pretend to be deaf
huanco rincri [hollow ears]	a disobedient person who does not hear what is ordered
ccaymarayak	one whose sense of taste is damaged
sonconak [one without a heart]	one who does not have the use of his reason
uparac [*upa:* deaf]	one who does not have the use of his reason because of youth or old age
quecapa	an idiot
utek	a fool
panra	a stupid person
ppanrayani	to become stupid
ppanrachanani	to become completely stupid

MODERN QUECHUA BODY TERMINOLOGY

The Body

External body parts

kurku	body
paña lado	right side
lloque lado	left side
ñawpa lado	front
quepa lado	back
ccara	skin

Head

uma	head

chuccha	hair
matti	forehead
uya	face
ñaui	eye
ñauirurun	crystalline lens
ñauiq chuyan	iris
llipi	eyelid
pillurqui	eyelash
quechifra	eyebrow
rinri	ear
lepo	external ear
rinric tocon	ear canal
senca	nose
sencac tocon	nostrils
uyapapa	cheek
kaqui	chin
simi	mouth
wirpa	upper lip
sirpa	lower lip
kallu	tongue
quiru	tooth
ñawpa quiru	incisor
alco kiru	canine
mara quiru	molar
huaque	wisdom tooth
quiru raqui	space between teeth
kallu	tongue
toncor	throat
toncor moco	Adam's apple
cunca	neck
muchu	back of neck

Torso

casco	chest
ñuñu	breast
huasa	back
wicsa	abdomen
pacca	lap
puputi	navel
siquipapa	buttocks
pullu	pubic hair
cacaran	clitoris
raca	vagina
ullu	penis
corota	testicle

Limbs

ricra	shoulder
challhuan	arm
huallhuacu	armpit
maquicuichun	forearm
maqui	hand
rucana	finger
chaca	leg
mocco	knee
chupa	thigh
chaquisenca	lower leg
chupa	calf
chaqui	foot
taycu	heel
chaquirucana	toe
chaquisaruna	sole

Internal body parts (skeletal)

tullu	bone
tucruru	skull
karmin	shoulder blade
huamani	flat part of shoulder blade
huasantullu	spine
huactatullu	rib
casacotullu	sternum
marccatullu	humerus
piruru	ulna
huachitullu	radius
maquitullucuna	bones of hand
tecnitullu	pelvis
wichu	femur
huamantullu	tibia
chaquitullucuna	bones of foot

Internal body parts (soft)

sirka	vein
yahuar	blood
yahuar puriynin	circulation of blood
anccu	nerve
ñoscon	brain
sankar	trachea
millputi	esophagus
amuclli	tonsils
machin	biceps
sonco	heart
sorquan	lung

wicsa	stomach
raccu chunchull	large intestine
ñañu chunchull	small intestine
huatan	appendix
rurun	kidney
succupin	pancreas
cucupin	liver
hayaque	gall bladder
quisma	uterus
sullu	fetus
tami	placenta
ispay puru	bladder
ispay	urine
yuma	semen

The Senses

cawllaycuna	the senses
cahua	sight
uyari	hearing
mushqui	smell
misquiq	taste
cullana	touch

NOTES

CHAPTER 1 INTRODUCTION

1. The term *Incas* will be used to refer to the ruling class of Tahuantinsuyu until the end of organized resistance in the late sixteenth century, and the term *Andeans* to refer to the people of the Central and Southern Andes in general, before and after the Conquest.

2. A history of social-scientific approaches to the body, and a bibliography of works in the field, is provided by J. M. Berthelot and M. Drulhe, eds., "Les sociologies et le corps," *Current Sociology* 33:2 (Summer 1985):1–200.

3. Mary Douglas, *Natural Symbols: Explorations in Cosmology* (New York: Pantheon, 1970).

4. Bryan Turner, *The Body and Society* (Oxford: Basil Blackwell, 1984).

5. For a discussion and criticism of this schism, see Brian Morris, *Anthropological Studies of Religion: An Introductory Text* (Cambridge: Cambridge University Press, 1987).

6. Sam Gill, *Native American Religious Action: A Performance Theory* (Columbia: University of South Carolina Press, 1987), p. 5.

7. Some interesting work has been done on the role of sensory perception, particularly sight and hearing, in different religious traditions. See, e.g., David Shane Chidester, "Word and Light: Perception and Symbolic Forms in the Augustinian Tradition" (Ph.D. diss., University of California, Santa Barbara, 1981); Diana Eck, *Darśan: Seeing the Divine Image in India,* rev. 2d ed. (Chambersburg, Penn.: Anima Books, 1985); C. Mackenzie Brown, "Purana as Scripture: From Sound to Image of the Holy Word in Hindu Tradition," *History of Religions* 26:1 (August 1988): 68–86; and Margaret Miles, *Image as Insight: Visual Understanding in Western Christianity and Secular Culture* (Boston: Beacon Press, 1985).

8. See Walter Ong, "World as View and World as Event," *American Anthropologist* 71 (1969):634–647. See also Constance Classen, "Creation by Light/Creation by Sound: A Sensory Analysis of Two South American Cosmologies," in *The Varieties of Sensory Experience: A Source Book in the Anthropology of the Senses,* ed. D. Howes (Toronto: University of Toronto Press, 1991), pp. 239–255.

9. David Howes, "The Shifting Sensorium: A Critique of the Textual Revolution in Anthropological Theory" (paper presented at the Twelfth International Congress of Anthropological and Ethnological Sciences, Zagreb, Yugoslavia, July 26, 1988).

10. See, e.g., Claude Lévi-Strauss, *The Raw and the Cooked: Introduction to a Science of Mythology,* vol. 1, trans. J. and D. Weightman (New York: Harper and Row, 1969), pp. 147–163.

11. Our extreme visualism can also cause us to distort our perception of our own culture by leading us to downplay the role other senses have had in our cultural development. See, e.g., Alain Corbin, *The Foul and the Fragrant: Odour and the French Social Imagination,* trans. M. Kochan, R. Porter, and C. Prendergast (Cambridge, Mass.: Harvard University Press, 1988).

12. Clifford Geertz, *The Interpretation of Cultures* (New York: Basic Books, 1973), p. 452.

13. See, e.g., David Howes, ed., *The Varieties of Sensory Experience: A Source Book in the Anthropology of the Senses* (Toronto: University of Toronto Press, 1991); Anthony Seeger, *Nature and Society in Central Brazil: The Suya Indians of Mato Grosso* (Cambridge, Mass.: Harvard University Press, 1981); Paul Stoller, *The Taste of Ethnographic Things: The Senses in Anthropology* (Philadelphia: University of Pennsylvania Press, 1989).

14. Howes, "Shifting Sensorium," p. 9. See further, David Howes, "Controlling Textuality: A Call for a Return to the Senses," *Anthropologica* 32:1 (1990):55–73.

15. George Urioste, ed., *Hijos de Pariya Qaqa: la tradición oral de Waru Chiri* [ca. 1608] vol. 2 (Syracuse, N.Y.: Syracuse University Maxwell School of Citizenship and Public Affairs, 1983), pp. 161–171.

16. Felipe Guaman Poma de Ayala, *El primer nueva corónica y buen gobierno* [ca. 1613], ed. J. Murra and R. Adorno, vol. 2 (México: Siglo Veintiuno, 1980), p. 781.

17. Ibid.

18. E.g., Geoffrey Conrad and Arthur Demarest, *Religion and Empire: The Dynamics of Aztec and Inca Expansionism* (Cambridge: Cambridge University Press, 1984). The ethnohistorian authors state that archaeology "belies" the claim of the Incas to have originated in the Titicaca Basin of the Southern Andes (p. 95). On one level this is similar to saying that archaeology "belies" the claim of Christians to have originated in the Garden of Eden.

19. Modern Andean myth holds to this view, also. See, e.g., Alejandro Ortiz Rescaniere, "El mito de la escuela," in *Ideología mesiánica del mundo andino*, ed. J. Ossio A. (Lima: Colección Biblioteca de Antropología, 1973), pp. 238–243.

20. Joseph Bastien, *Mountain of the Condor: Metaphor and Ritual in an Andean Ayllu* (St. Paul, Minn.: West, 1978), and *Healers of the Andes* (Salt Lake City: University of Utah Press, 1987).

21. E.g., Charlene and Ralph Bolton, "Rites of Retribution and Restoration in Canchis," *Journal of Latin American Lore* 2:1 (1976):97–114; Douglas Sharon, *Wizard of the Four Winds: A Shaman's Story* (New York: Free Press, 1978); and Donald Joralemon, "The Role of Hallucinogenic Drugs and Sensory Stimuli in Peruvian Ritual Healing," *Culture, Medicine and Pyschiatry* 8 (1984):399–430.

22. Veronica Cereceda, "Aproximaciones a una estética andina: de la belleza al tinku," in *Tres reflexiones sobre el pensamiento andino*, ed. J. Medina (La Paz: Hisbol, 1987), pp. 133–231; Regina Harrison, *Signs, Songs, and Memory in the Andes: Translating Quechua Language and Culture* (Austin: University of Texas Press, 1989).

23. See, e.g., R. T. Zuidema, "Masks in the Incaic Solstice and Equinoctial Rituals," in *The Power of Symbols: Masks and Masquerade in the Americas*, ed. N. R. Crumrine and M. Halpin (Vancouver: University of British Columbia Press, 1983), pp. 149–156.

24. Billie Jean Isbell, "The Metaphoric Process: 'From Culture to Nature and Back Again'," in *Animal Myths and Metaphors in South America*, ed. G. Urton (Salt Lake City: University of Utah Press, 1985), pp. 285–313; Lawrence Sullivan, "Sound and Senses: Toward a Hermeneutics of Performance," *History of Religions* 26:1 (1986):1–33.

25. Alfredo López Austin, for example, has examined how the Nahua Indians of Mexico conceived of the human body. Although his work deals more

with Nahua concepts of the constitution and functioning of the human body, and the social consequences of those concepts, than with the cosmological role of the body, it nonetheless provides an interesting comparison with the analysis of Inca cosmology and the body undertaken here. See Alfredo López Austin, *The Human Body and Ideology: Concepts of the Ancient Nahuas*, 2 vols., trans. T. and M. Ortiz de Montellano (Salt Lake City: University of Utah Press, 1988).

26. See, for instance, Leonard Barkan, *Nature's Work of Art: The Human Body as Image of the World* (New Haven, Conn.: Yale University Press, 1975).

27. Tzetan Todorov, *The Conquest of America: The Question of the Other*, trans. R. Howard (New York: Harper and Row, 1984), p. 4.

CHAPTER 2 COSMOLOGY

1. John Murra discusses this system of exchange in "El 'control vertical' de un máximo de pisos ecológicos en la economía de las sociedades andinas," in *Visita de la Provincia de León de Huanuco*, ed. Ortiz de Zúñiga, vol. 1 (Huanuco, Peru: Universidad Hermilio Valdizan, 1972), pp. 383–406.

2. Billie Jean Isbell, *To Defend Ourselves: Ecology and Ritual in an Andean Village* (Austin: Institute of Latin American Studies and University of Texas Press, 1978), p. 203.

3. Enrique Oblitas Poblete, *Cultura Callawaya* (La Paz: Talleres Gráficos Bolivianos, 1963), p. 50.

4. June Nash, *We Eat the Mines and the Mines Eat Us: Dependency and Exploitation in Bolivian Tin Mines* (New York: Columbia University Press, 1979), p. 124.

Bernard Mishkin writes of a community in the Peruvian Andes: "Against 'frost,' bonfires are started in the fields. However, it should be noted that many Indians consider the bonfire (cconuy) as a means of engaging the 'frost' in physical combat and not of protecting the plants by raising the temperature in the vicinity" ("The Contemporary Quechua," in *Handbook of South American Indians*, vol. 2, ed. J. H. Steward [New York: Cooper Square, 1946], p. 425).

5. The verb *aynini* means "to recompense." Diego de González Holguín, *Vocabulario de la lengua general de todo el Perú* [1608] (Lima: Imprenta Santa Maria, 1952). All definitions of Quechua terms are from this work unless otherwise stated.

6. John Earls and Irene Silverblatt, "La realidad física y social en la cosmología andina," in *Actes du XLlle Congrès International des Américanistes*, vol. 4 (Paris, 1976), p. 309.

7. Ernst Cassirer writes: that "it is eminently the differentiation of the parts of his own body that serves man as a basis for all other spatial specifications" (*The Philosophy of Symbolic Forms*, vol. 1: *Language*, trans. R. Mannheim [New Haven, Conn.: Yale University Press, 1953], p. 206).

8. Domingo de Santo Tomás, *Léxicon o vocabulario de la lengua general del Perú* [1560] (Lima: Universidad Nacional Mayor de San Marcos, 1951). Similar terms are noted in the modern Andes by Louisa Stark, "The Lexical Structure of Quechua Body Parts," *Anthropological Linguistics* 2 (January 1969):5.

9. Luis Capoche, "Relación general de la villa imperial de Potosí" [1585], in *Relaciones histórico-literarias de la América meridional*, ed. L. Hanke (Madrid: Ediciones Atlas, 1959), pp. 139–140.

10. Tristan Platt, "Mirrors and Maize: The Concept of *Yanantin* Among the Macha of Bolivia," in *Anthropological History of Andean Polities*, ed. J. Murra, N. Wachtel, and J. Revel (Cambridge: Cambridge University Press, 1986), pp. 245–246.

11. See, e.g., Joan de Santacruz Pachacuti Yamqui, "Relación de antigüedades deste reyno del Pirú" [ca. 1613], in *Tres relaciones de antigüedades peruanas,* ed. M. Jimenez de la Espada (Asunción, Paraguay: Editorial Guaraní, 1950), p. 268; Guaman Poma, *El primer nueva corónica,* vol. 1, p. 96.

12. Cassirer discusses the close relationship between spatial and temporal terms in many cultures in *Symbolic Forms,* vol. 1, pp. 215–217.

13. Martín de Murúa, *Historia del origen y genealogía real de los reyes Incas del Perú* [1590], ed. C. Bayle (Madrid: Instituto Santo Toribio de Mogrovejo, 1946), p. 211.

14. Stark, "Quechua Body Parts," pp. 5, 10.

15. Gary Urton writes that in the Andes constellations are made up not only of stars, but also of the black spaces between stars. See his *At the Crossroads of the Earth and the Sky: An Andean Cosmology* (Austin: University of Texas Press, 1981), pp. 9–10.

16. Bernabé Cobo, "Historia del Nuevo Mundo" [1653], in *Obras del P. Bernabé Cobo,* vol. 2, ed. F. Mateos (Madrid: Ediciones Atlas, 1964), p. 194.

17. Pachacuti Yamqui, "Relación," p. 271.

18. Juan Polo de Ondegardo, "Instrución contra las ceremonias y ritos que usan los indios conforme al tiempo de su infidelidad" [1567] in *Colección de libros y documentos referentes a la historia del Perú,* ed. H. H. Urteaga and C. A. Romero (Lima: Sanmartí, 1916), p. 198; Ana María Mariscotti de Gorlitz, "Los curi y el rayo," in *Actes du XLlle Congrès International des Américanistes,* vol. 4, (Paris, 1976), pp. 365–376.

19. Bastien, *Healers,* pp. 46, 72.

20. Ibid., p. 69.

21. Ibid., p. 74.

22. Glynn Custred writes that "most of the contexts of *songo* found in the colonial Quechua literature . . . indicate belief in a literal merger of mind, temperament and emotion with the corporal organs of the heart and the stomach." ("Inca Concepts of Soul and Spirit," in *Essays in Humanistic Anthropology: A Festschrift in Honor of David Birney,* ed. B. T. Grindal and D. M. Warren [Washington, D.C.: University Press of America, 1979], p. 286).

23. In the *talegas* (rectangular woven bags,) of the Aymara of the Chilean Andes, we have a modern Andean stylistic representation of the body. Each side of the bag has vertical stripes of different colors that are repeated so that the left half of the side mirrors the right half, with one odd stripe situated in the center. The bag is known as a "body"; the stripes along the edge of each side, communicators with the outside world, are called "mouths"; and the center stripe is called a "heart." See Veronica Cereceda, "The Semiology of Andean Textiles: The Talegas of Isluga," in *Anthropological History of Andean Polities,* ed. J. Murra, N. Wachtel, and J. Revel (Cambridge: Cambridge University Press, 1986), pp. 159–161.

24. González Holguín, *Vocabulario,* p. 63.

25. Ibid., p. 65.

26. Ludovico Bertonio, *Vocabulario de la lengua aymara,* vol. 1, [1612] (Cochabamba, Bolivia: Ediciones Ceres, 1984), p. 424.

27. See, e.g., Louise Vinge, *The Five Senses: Studies in a Literary Tradition* (Lund, Sweden: Royal Society of Letters at Lund, 1975).

28. Polo de Ondegardo, "Instrución," p. 199.

29. Pachacuti Yamqui, "Relación," p. 220. Translated from Quechua into

Spanish by S. A. Lafone Quevedo in "El culto de Tonapa," in ibid., pp. 306–307; my translation into English.

30. Harrison, *Signs, Songs, and Memory,* pp. 78–82.

31. Andrew Miracle and Juan Yapita Moya, "Time and Space in Aymara," in *The Aymara Language in its Social and Cultural Context,* ed. M. J. Hardman (Gainesville: University of Florida Press, 1981), p. 33.

32. See, e.g., Pachacuti Yamqui, "Relación," p. 259; Polo de Ondegardo, "Instrución," p. 197.

33. Urton, *At the Crossroads,* pp. 17–18.

34. Claude Lévi-Strauss has examined how sound is used in similar ways in the Amazon; see his *From Honey to Ashes: Introduction to a Science of Mythology,* trans. J. and D. Weightman (New York: Harper and Row, 1973), pp. 361–422.

35. Billie Jean Isbell, "'Those Who Love Me': An Analysis of Andean Kinship and Reciprocity Within a Ritual Context," in *Andean Kinship and Marriage,* ed. R. Bolton and F. Mayer (Washington, D.C.: American Anthropological Association, 1977), p. 88.

As this reference makes evident, the eye has certain natural associations with fluidity, as well as with structure, in the Andes. The eye can be said to symbolize the production of structure from fluidity. This symbolism occurs in similar contexts in Andean mythology, as in the Viracocha cosmogony in which the sun and moon arise out of a lake (see Chapter 3).

36. R. T. Zuidema, "Inka Dynasty and Irrigation: Another Look at Andean Concepts of History," in *Anthropological History of Andean Polities,* ed. J. Murra, N. Wachtel, and J. Revel (Cambridge: Cambridge University Press, 1986), p. 193.

37. The Andean chroniclers of the sixteenth and seventeenth centuries display an interesting combination of sight and hearing in their language, as in the following examples: "A great bloodbath was heard" and "raising his eyes to heaven in the language of that land," from Pachacuti Yamqui, "Relación," pp. 250, 213; and "what they say we cannot see well," from Diego de Castro Titu Cusi Yupanqui, *Ynstrución del Ynga . . .* [1570], ed. Luis Millones (Lima: Ediciones El Virrey, 1985), p. 26.

38. Pachacuti Yamqui was the cacique, or chief, of a village to the south of Cuzco.

39. In the present-day Andes this cat is known as *ccoa.* Mishkin describes it as casting hail from its eyes and ears and serving as a sponsor to sorcerors ("Contemporary Quechua," p. 464).

40. This is written over the top of the original diagram but not reproduced in the figure presented here. Pachacuti Yamqui, "Relación," p. 227 n. 2.

41. Ibid.

42. The Spanish destroyed one that was about 5.5 meters high and represented a man and a women joined at the back, the man facing the east and the woman the west. See Pablo José de Arriaga, *The Extirpation of Idolatry in Peru* [1621], trans. and ed. L. Clark Keating (Lexington: University of Kentucky Press, 1968), p. 79.

43. R. T. Zuidema, "The Inka Kinship System: A New Theoretical View," in *Andean Kinship and Marriage,* ed. R. Bolton and F. Mayer (Washington, D.C.: American Anthropological Association, 1977), p. 264.

44. Irene Silverblatt, *Moon, Sun, and Witches: Gender Ideologies and Class in Inca and Colonial Peru* (Princeton, N.J.: Princeton University Press, 1987), p. 54.

45. Ibid., pp. 119–120.

46. Another possible interpretation is that the Earth as ground and interior is female, but as world male. See Isbell, *To Defend Ourselves,* p. 209.

47. Isbell sees symbols of state and power on the masculine side, and family and nurture on the feminine side. See ibid., p. 210.

48. As does Robert Hertz, for example, in his seminal paper "The Pre-eminence of the Right Hand: A Study in Religious Polarity," in *Right and Left,* ed. R. Needham (Chicago: University of Chicago Press, 1973), pp. 3–31.

49. Gary Urton and R. T. Zuidema, "La constelación de la Llama en los Andes peruanos," *Allpanchis* 9 (1976):59–119.

50. Olivia Harris, "Complementarity and Conflict: An Andean View of Women and Men," in *Sex and Age as Principles of Social Differentiation,* ed. J. S. LaFontaine (London: Academic Press, 1978), p. 28.

51. Another possible opposition is that of the Eyes of All Things on the left, and the destructive, hail-casting eyes of the choquechinchay cat on the right.

52. Henrique-Osvaldo Urbano, *Wiracocha y Ayar: héroes y funciones en las sociedades andinas* (Cusco: Centro de Estudios Rurales Andinos Bartolomé de Las Casas, 1981), pp. xxxviii–xxxix.

53. Murúa, *Origen y genealogía,* pp. 285, 343.

54. In 1621 a Spanish priest wrote that, after their stone huacas, what the Andeans revered most highly were their mallquis, "the bones or mummies of their pagan ancestors" (Arriaga, *Extirpation of Idolatry,* p. 27).

55. Pachacuti Yamqui, "Relación," pp. 217–218, 239.

56. Earls and Silverblatt, "Realidad," p. 319.

CHAPTER 3 CREATION

1. Pedro Sarmiento de Gamboa, "Historia Índica" [1572] in *Obras de Garcilaso de la Vega,* vol. 4, ed. C. Sáenz de Santa María (Madrid: Ediciones Atlas, 1965), pp. 207–210.

2. Pachacuti Yamqui, for example, writes of Viracocha: "Might not this man be the glorious apostle Saint Thomas?" ("Relación," p. 211).

3. Franklin Pease, *El dios creador andino* (Lima: Mosca Azul, 1973), p. 14.

4. John H. Rowe, "Inca Culture at the Time of the Spanish Conquest," in *Handbook of South American Indians,* vol. 2: *The Andean Civilizations,* ed. J. H. Steward (New York: Cooper Square, 1946), pp. 323–325. See also María Rostworowski de Díez Canseco, *Pesos y medidas en el Perú pre-hispánico* (Lima: Impresa Minerva, 1960).

5. Pachacuti Yamqui, "Relación," p. 210.

6. Cristóbal de Molina "of Cuzco," "Fábulas y ritos de los Incas" [1574], in *Las crónicas de los Molinas,* ed. F. Loayza (Lima: Los pequeños grandes libros de la historia americana, 1943), p. 20.

7. Silverblatt, *Moon, Sun, and Witches,* p. 19.

8. Cobo, "Nuevo Mundo," p. 150.

9. Molina of Cuzco, "Fábulas y ritos," p. 14.

10. Garcilaso de la Vega, *Comentarios reales de los Incas* [1609], vol. 1, ed. A. Rosenblat (Buenos Aires: Emecé, 1945), p. 79.

11. Pierre Duviols, "Un symbolisme andin du double, la lithomorphose de l'ancêtre," in *Actes du XLlle Congrès International des Américanistes* (Paris, 1976), pp. 359–364.

12. Cobo, "Nuevo Mundo," p. 151.

13. Arriaga, *Extirpation of Idolatry,* p. 24.

14. Ibid. Cassirer notes that "for mythical thinking the relation between what a thing 'is' and the place in which it is situated is never purely external and accidental; the place is itself a part of the thing's being, and the place confers very specific inner ties upon the thing" (*Symbolic Forms,* vol. 2, p. 92).

15. Hernando de Santillan, "Relación del origen, descendencia, política y gobierno de los Incas" [1563], in *Crónicas peruanas de interés indígena,* ed. F. Esteve Barba (Madrid: Ediciones Atlas, 1968), p. 133.

16. Cobo, "Nuevo Mundo," p. 151.

17. See, e.g., Bastien, *Mountain of the Condor,* p. 97.

18. Isbell, *To Defend Ourselves,* p. 71.

19. Sarmiento de Gamboa, "Historia Índica," p. 209.

20. Garcilaso de la Vega, *Comentarios reales,* p. 106.

21. R. T. Zuidema looks at how designs on certain Inca textiles may contain calendrical information in "The Inca Calendar," in *Native American Astronomy,* ed. A. F. Aveni (Austin: University of Texas Press, 1977), pp. 221–226. Gail Silverman-Proust examines the iconography of modern Andean textiles in "Weaving Technique and the Registration of Knowledge in the Cuzco Area of Peru," *Journal of Latin American Lore* 14:2 (1988):207–241.

22. Inca priests also dressed in white, affirming the association of white with sacrality; Murúa, *Origen y genealogía,* p. 257.

23. See, e.g., Guaman Poma, *El primer nueva corónica,* vol. 3, p. 854.

24. Sharon, *Wizard of the Four Winds,* p. 77.

25. Urioste, *Hijos de Pariya Qaqa,* vol. 1, p. 177.

26. Pachacuti Yamqui, "Relación," p. 211.

27. Ibid., p. 212.

28. Molina of Cuzco, "Fábulas y ritos," p. 30.

29. Pedro de Cieza de León, *El señorío de los Incas* [1550], ed. M. Ballesteros (Madrid: Historia 16, 1985), p. 36.

30. Molina of Cuzco, "Fábulas y ritos," pp. 13–14.

31. Pachacuti Yamqui, "Relación," p. 215.

32. Cobo, "Nuevo Mundo," p. 150.

33. Urioste, *Hijos de Pariya Qaqa,* vol. 2, p. 181.

34. Cobo, "Nuevo Mundo," p. 151.

35. See Catherine Allen Wagner, "Coca, Chicha and Trago: Private and Communal Rituals in a Quechua Community" (Ph.D. diss., University of Illinois, Urbana-Champaign, 1978), p. 47; and Nathan Wachtel, *The Vision of the Vanquished: The Spanish Conquest of Peru Through Indian Eyes, 1530–1570,* trans. B. and S. Reynolds (New York: Barnes and Noble, 1977), p. 40.

36. Harrison, *Signs, Songs, and Memory,* pp. 76–77.

37. Pachacuti Yamqui, "Relación," p. 211.

38. González Holguín, *Vocabulario,* p. 297.

39. Pachacuti Yamqui, "Relación," p. 211.

40. Ibid.

41. R. T. Zuidema, *The Ceque System of Cuzco* (Leiden: E. J. Brill, 1964), p. 169.

42. Urioste, *Hijos de Pariya Qaqa,* vol. 1, p. 19; R. T. Zuidema, "Shafttombs and the Inca Empire," *Journal of the Steward Anthropological Society* 9 (1978):151.

43. Cobo, "Nuevo Mundo," p. 150.

44. Cieza, *El señorío de los Incas,* p. 36.

45. Urioste, *Hijos de Pariya Qaqa,* vol. 2, p. 181.

46. See, e.g., Juan de Betanzos, "Suma y narración de los Incas" [1551], in

Crónicas peruanas de interés indígena, ed. F. Esteve Barba, (Madrid: Ediciones Atlas, 1968), p. 9.

CHAPTER 4 MYTHS OF THE INCAS

1. Condensed from Garcilaso de la Vega, *Comentarios reales,* vol. 1, pp. 40–44.

2. José María Arguedas, *Yawar Fiesta,* trans. F. Horning Barraclough (Austin: University of Texas Press, 1985), pp. 164–165.

3. Sarmiento de Gamboa, "Historica Índica," p. 217.

4. Olivia Harris writes that, while man and woman form an archetypal pair of complementary opposites in the Andes, older and younger brother form an archetypal pair of antagonistic similarity; see "Complementarity and Conflict," p. 36. Many Andean myths and stories, in fact, deal with the struggle between two brothers.

5. Molina of Cuzco, "Fábulas y ritos," p. 61. Such a ritual battle regularly takes place between the two moieties of many modern Andean communities; see, e.g., Harris, "Complementarity and Conflict," p. 36.

6. Pachacuti Yamqui, "Relación," pp. 214–217.

7. Gary Urton examines the social history surrounding this mythical birthplace of the Incas in *The History of a Myth: Pacariqtambo and the Origin of the Inkas* (Austin: University of Texas Press, 1990).

8. Pachacuti Yamqui, pp. 213–219.

9. Garcilaso de la Vega, *Comentarios reales,* vol. 1, p. 46.

10. José María Arguedas, "Taki Parwa y la poesía quechua de la república" *Letras Peruanas* 12:3 (1955):74.

11. Pachacuti Yamqui, "Relación," p. 215.

12. Molina of Cuzco, "Fábulas y ritos," p. 35.

13. Cieza, *El señorío de los Incas,* p. 42.

14. From Conrad and Demarest, *Religion and Empire,* p. 98. This list is intended to serve solely as a guide to the reader and not to place Inca mythology within a historical framework.

15. Condensed from Betanzos, "Suma y narración," pp. 15–56.

16. Cieza, *El señorío de los Incas,* p. 124.

17. Ibid., pp. 127, 136.

18. Ibid., pp. 137–138.

19. See, e.g., Pachacuti Yamqui, "Relación," p. 238.

20. Sarmiento de Gamboa, "Historia Índica," p. 236.

21. Martín de Murúa, *Historia general del Perú: origen y descendencia de los Incas* [1613], ed. M. Ballesteros-Gaibrois (Madrid: Instituto Gonzalo Fernández de Oviedo, 1962), p. 88.

22. Titu Cusi Yupanqui, *Ynstrución,* p. 7.

23. Pachacuti Yamqui, "Relación," p. 238.

24. Molina of Cuzco, "Fábulas y ritos," p. 20.

25. Sarmiento de Gamboa, "Historia Índica," p. 237.

26. Cobo, "Nuevo Mundo," p. 78.

27. Sarmiento de Gamboa, "Historia Índica," p. 237. Illapa was personified by the Incas as a man in the sky; see Juan Polo de Ondegardo, "Los errores y supersticiones de los Incas . . . ," in *Colección de libros y documentos referentes a la historia del Perú,* ser. 1, vol. 3, ed. H. H. Urteaga and C. A. Romero (Lima: Sanmartí, 1916), p. 6.

28. R. T. Zuidema, "Myth and History in Ancient Peru," in *The Logic of Culture: Advances in Structural Theory and Methods,* ed. I. Rossi and contributors (London: Tavistock, 1982), p. 157.

29. Pachacuti Yamqui, "Relación," p. 228.

30. Garcilaso de la Vega, *Comentarios reales,* vol. 1, p. 263.

31. Ibid., vol. 1, p. 218.

32. Ibid., vol. 2, p. 168.

33. Ibid., vol. 1, p. 265.

34. Sarmiento de Gamboa, "Historia Índica," p. 238.

35. Garcilaso de la Vega, *Comentarios reales,* vol. 1, p. 47.

36. Urbano, *Wiracocha y Ayar,* p. xlix.

37. Catherine Allen, "Patterned Time: The Mythic History of a Peruvian Community," *Journal of Latin American Lore* 10:2 (1984):155.

38. The inhabitants of the Sibundoy Valley in the Colombian Andes, in a creation myth that has similarities with the Viracocha cosmogony, describe the first age of the world as characterized by darkness, the second age by light, the third age by raw food, and the fourth by cooking. See John Holmes McDowell, *Sayings of the Ancestors: The Spiritual Life of the Sibundoy Indians* (Lexington: University Press of Kentucky, 1989). For an analysis of cooking as a symbolic transformation of nature into culture, see Lévi-Strauss, *The Raw and the Cooked.*

39. Sarmiento de Gamboa, "Historia Índica," p. 216.

40. Pachacuti Yamqui, "Relación," p. 215.

41. Ibid.

42. Ibid., p. 238.

43. Ibid., pp. 213–214.

44. Cristóbal de Vaca, "Discurso sobre la descendencia y gobierno de los Incas" [1542], in *Colección de libros y documentos referentes a la historia del Perú,* ser. 2, vol. 3, ed. H. Urteaga (Lima: Sanmartí, 1920), p. 7.

45. Garcilaso de la Vega, *Comentarios reales,* vol. 1, p. 114.

46. Ibid., vol. 1, p. 64.

47. On this, see, e.g., Gary Gossen, *Chamulas in the World of the Sun: Time and Space in a Maya Oral Tradition* (Cambridge, Mass.: Harvard University Press, 1974).

48. Condensed from Pachacuti Yamqui, "Relación," pp. 229–230.

49. The assertion that the voice of the Creator can destroy the "light of the Sun," the Inca, contained in the Huarochiri myth, may represent a local response to this.

50. Molina of Cuzco, "Fábulas y ritos," p. 21.

51. Andean holy objects and places were and are oracles that communicate with their followers through speech.

52. Betanzos, "Suma y narración,"p. 50.

53. Ibid., p. 52.

54. Pachacuti Yamqui, "Relación," p. 230.

55. That the Incas recognized that sight and light have their limits, however, is evidenced by the reflection attributed to Pachacuti that the sun, which is so easily blocked by a cloud, cannot be the all-powerful Creator; see Cobo, "Nuevo Mundo," pp. 78–79.

56. Murúa, *Origen y genealogía,* pp. 105–106.

57. Cieza, *El señorío de los Incas,* p. 140.

58. Betanzos, "Suma y narración," pp. 51–52.

59. Garcilaso de la Vega, *Comentarios reales,* vol. 1, p. 265.

CHAPTER 5 RITUALS AND INSTITUTIONS

1. Polo de Ondegardo, "Errores y supersticiones," p. 38.
2. Betanzos, "Suma y narración," p. 23.
3. Ibid., p. 52.
4. Murúa, *Origen y genealogía,* p. 355.
5. Sarmiento de Gamboa, "Historia Índica," p. 238.
6. Garcilaso de la Vega, *Comentarios reales,* vol. 1, p. 228. In modern Andean tradition this conception of agriculture as intrusive persists. Many Andeans, for instance, will not work the Earth on certain days when she is thought to be particularly sensitive, or "alive." See Bernabé Condori and Rosalind Gow, *Kay Pacha* (Cusco: Centro de Estudios Rurales Andinos Bartolomé de Las Casas, 1982), p. 5.
7. Polo de Ondegardo writes that the Andeans did not think that sexual relations between a single man and woman were a sin, see his "Instrución," p. 203.
8. The *yanacona* (blacks), male servants of the Incas, possibly formed the male counterpart of the acllas. They were selected for their superior physiques from among the sons of the *curacas* (chiefs) of Tahuantinsuyu. Once they entered the service of the Incas they came under the direct authority of the Inca and were no longer subject to their local regulations. Pachacuti Yamqui implies that the yanacona were kept celibate in their youth ("Relación," p. 224). As adults, however, they were apparently allowed to marry, and, in fact, the Inca often presented them with acllas who were not intended for religious service, for wives. See Zuidema, *Ceque System of Cuzco,* p. 225.
9. Murúa, *Origen y genealogía,* p. 387; Garcilaso de la Vega, *Comentarios reales,* vol. 1, pp. 184–192. Some chroniclers suggest that the Inca had sexual access to all the acllas; e.g., Cobo, "Nuevo Mundo," p. 232. If so, it could only be as the incarnation of the Sun, the deity to which the acllas were mated.
10. Silverblatt, *Moon, Sun, and Witches,* p. 107.
11. Murúa, *Origen y genealogía,* p. 269. Anonymous, "Relación de las costumbres antiguas de los naturales del Pirú" [ca. 1600], in *Crónicas peruanas de interés indígena,* ed. F. Esteve Barba (Madrid: Ediciones Atlas, 1968), p. 158.
12. Cobo, "Nuevo Mundo," p. 248.
13. Betanzos, "Suma y narración," p. 55.
14. Garcilaso de la Vega, *Comentarios reales,* vol. 2, p. 53.
15. Anonymous, "Costumbres antiguas," p. 158.
16. Cobo, "Nuevo Mundo," p. 145.
17. Ibid., pp. 247–250; Murúa, *Origen y genealogía,* 244–245.
18. Guaman Poma, *El primer nueva corónica,* vol. 1, p. 177.
19. Harris writes of the Laymis, a modern community of Bolivian Aymara, that "it is considered extremely unfortunate to die without having been married, and the corpse has to be interred with a symbolic companion for the after life: a hen for a man and a cock for a woman. Unmarried people are thought to have a far harder journey to the land of the dead. They can travel only at night. . . . The road unmarried souls must go is over thorn bushes, while those of married people travel on paths, and by day" ("Complementarity and Conflict," p. 28).
20. See, e.g., Murúa, *Origen y genealogía,* p. 354.
21. Ibid.; Cieza, *El señorío de los Incas,* pp. 93–94.
22. Pachacuti Yamqui, "Relación," pp. 242–243, 247.
23. Molina of Cuzco, "Fábulas y ritos," p. 67; Cobo, "Nuevo Mundo," pp. 245–246.

24. Molina of Cuzco, "Fábulos y ritos," p. 67; Cobo, "Nuevo Mundo," p. 246.

25. Molina of Cuzco, "Fábulos y ritos," pp. 48–60; Garcilaso de la Vega, *Comentarios reales,* vol. 2, pp. 55–62.

26. Molina of Cuzco, "Fábulos y ritos," p. 64; Cobo, "Nuevo Mundo," p. 247.

27. Titu Cusi Yupanqui, *Ynstrución,* p. 18.

28. For information on Inca age grades, see John H. Rowe, "The Age-Grades of the Inca Census," in *Miscellanea Paul Rivet,* vol. 2, pp. 499–522 (Mexico: Trigésimo primero Congreso Internacional de Americanistas, Universidad Autónoma de México, 1958).

29. Guaman Poma, *El primer nueva corónica,* vol. 1, pp. 183, 203.

30. Ibid., vol. 1, pp. 171, 190.

31. Zuidema, *Ceque System of Cuzco,* pp. 213–235.

32. Molina of Cuzco, "Fábulas y ritos," pp. 69–70, 75–77; Betanzos, "Suma y narración," p. 32.

33. Rodrigo Hernández Príncipe, "Visitas de Rodrigo Hernández Príncipe [1621–1622]," in *Cultura andina y represion . . . ,* ed. P. Duviols (Cuzco: Centro de Estudios Rurales Andinos Bartolomé de Las Casas, 1986), pp. 472–473. The meaning of this ritual is examined by Pierre Duviols in "La capacocha," *Allpanchis* 9 (1976):11–54.

34. Guaman Poma, *El primer nueva corónica,* vol. 1, p. 236.

35. Molina of Cuzco, "Fábulas y ritos," p. 73; Betanzos, "Suma y narración," p. 32.

36. Thérèse Bouysse-Cassagne, "Urco and Uma: Aymara Concepts of Space," in *Anthropological History of Andean Polities,* ed. J. Murra, N. Wachtel, and J. Revel (Cambridge: Cambridge University Press, 1986), p. 214.

37. Cobo, "Nuevo Mundo," pp. 224–225; Arriaga, *Extirpation of Idolatry,* pp. 32–41; Anonymous, "Costumbres antiguas," pp. 161–169. This last writer, a Spanish priest, gives a lengthy and highly interesting account of the religious organization of the Incas. Unfortunately, his description is too penetrated with Christian concepts to be very reliable.

38. Murúa, *Origen y genealogía,* p. 156.

39. Polo de Ondegardo, "Errores y supersticiones," p. 29.

40. Anonymous, "Costumbres antiguas," p. 169.

41. Ibid.

42. For an elaboration of this theme, see Constance Classen, "Aesthetics and Asceticism in Inca Religion," *Anthropologica* 32:1 (1990):101–106.

43. Molina of Cuzco "Fábulas y ritos," pp. 71–72.

44. Cobo, "Nuevo Mundo," pp. 167–186. Zuidema discusses the relationship between the ceques and the Inca social organization in *The Ceque System of Cuzco.*

45. Cobo, "Nuevo Mundo," p. 164.

46. Zuidema, "Inka Dynasty and Irrigation," p. 187.

47. Cobo, "Nuevo Mundo," p. 174.

48. Zuidema, "Inka Dynasty and Irrigation," p. 189.

49. Betanzos, "Suma y narración," p. 32.

50. Zuidema, "Masks," p. 151.

51. Murúa, *Historia general,* pp. 239–240. For similar reasons, defeated enemy troops were also not allowed to look up at the sky. The inference is made here that eye contact with the sky is an important prerequisite for communication with the deities who reside in the sky. See ibid., p. 116.

52. Ibid., p. 240.

53. Pedro Pizarro, for example, writes that "a very fine, thin mantle through which one could see was held up by two women before [Inca Atahualpa], and they covered him up with it so that no one should see him, for it is the custom of some of these lords not to be seen save rarely by their vassals" (*Relation of the Discovery and Conquest of the Kingdoms of Peru* [1571], vol. 1, ed. and trans. P. A. Means (New York: Cortés Society, 1921), p. 175).

54. Murúa, *Origen y genealogía,* p. 117; Arriaga, *Extirpation of Idolatry,* p. 51; Betanzos, "Suma y narración," p. 32. In the present-day Andes it is still believed dangerous to look at sacred beings; see Harry Tschopik, Jr., "The Aymara of Chucuito, Peru," *Anthropological Papers of the American Museum of Natural History* 44 (1951):133–308.

55. Rowe, "Inca Culture," p. 300.

56. Guaman Poma, *El primer nueva corónica,* vol. 1, p. 229.

57. Zuidema writes that "at the end of their initiation ceremonies the noble youths went out to a *chacra,* a cultivated field, near a lake, to have their ears pierced. Clearly the act was intended to inaugurate their 'fertility,' and closeness to water was a necessary prerequisite" ("Inka Dynasty and Irrigation," p. 193).

58. In the early Andean period nose and lip ornaments, as well as ear ornaments were common in the Andes. However, by the time of the Incas only ear ornaments were used. See Wendell C. Bennet, "The Andean Highlands," in *Handbook of South American Indians,* vol. 2: *The Andean Civilizations* (New York: Cooper Square, 1963), p. 31. Anthony Seeger looks at the sensory implications of body ornaments in "The Meaning of Body Ornaments: A Suya Example," *Ethnology* 14 (1975):211–224.

59. Guaman Poma says that Tupac Inca was called Viracocha for his ability to speak with the huacas; his power of speech thereby allied him with the supreme mediator of sound (*El primer nueva corónica,* vol. 1, p. 234). According to Hernando de Santillan, Tupac Inca was also said to have spoken while still in the womb ("Relación," p. 111).

60. Cieza, *El señorío de los Incas,* pp. 118–119.

61. Guaman Poma describes Inca Roca as a "great talker" and one who "spoke with thunder" (*El primer nueva corónica,* vol. 1, p. 83).

62. Pachacuti Yamqui, "Relación," p. 224.

63. Murúa, *Origen y genealogía,* 114.

64. Guaman Poma, *El primer nueva corónica,* vol. 1, pp. 143, 287, 308.

65. Cristóbal Molina "el Almagrista," "Relación de muchas cosas acaescidas en el Perú" [1572], in *Crónicas peruanas de interés indígena,* ed. F. Esteve Barba. (Madrid: Ediciones Atlas, 1968), p. 82.

66. Urton, *At the Crossroads,* p. 32.

67. Garcilaso de la Vega, *Comentarios reales,* vol. 2, p. 62.

68. For example, in the rite of citua; see Cobo, "Nuevo Mundo," pp. 218–219.

69. Polo de Ondegardo, "Instrución," p. 190.

70. Garcilaso de la Vega, *Comentarios reales,* vol. 2, pp. 58–59.

71. González Holguín, *Vocabulario,* p. 156.

72. Charlene and Ralph Bolton have noted that in modern Andean rituals sweet flavors are used to attract and bitter flavors to repel ("Retribution and Restoration," pp. 107–111). In Inca ritual, flavors could presumably have been used in similar, and probably more complex, ways.

73. Murúa, *Origen y genealogía,* pp. 117, 267; Anonymous, "Costumbres antiguas," p. 155.

74. Murúa, *Historia general,* p. 238.

75. Molina el Almagrista, "Relación," p. 75.

76. Murúa, *Origen y genealogía,* pp. 262–263.

77. Although on one level the acllas were associated with the precultural Earth, on another they were the guardians of Inca esthetic values. The acllas were chosen for their exceptional beauty, they wove the finest cloth in the empire, prepared the finest chicha and foods, and had the finest singing voices. For a description of the acllas' excellencies, see, e.g., Anonymous, "Costumbres antiguas," pp. 171–173; and Guaman Poma, *El primer nueva corónica,* vol. 1, pp. 274–275.

78. Pizarro, *Discovery and Conquest,* vol. 1, p. 225.

79. Garcilaso de la Vega, *Comentarios reales,* vol. 1, pp. 200–201.

80. Ibid., vol. 2, pp. 58–59.

81. Pizarro, *Discovery and Conquest,* vol. 1, pp. 222–223.

82. Molina of Cuzco, "Fábulas y ritos," p. 51.

83. Ibid., pp. 62–63.

84. Guaman Poma, *El primer nueva corónica,* vol. 1, p. 287.

85. Ibid., vol. 1, p. 256.

86. Arriaga, *Extirpation of Idolatry,* p. 37.

87. Murúa, *Origen y genealogía,* p. 156. The anonymous chronicler also describes these ascetic diviners as spending their time contemplating the luminaries ("Costumbres antiguas," p. 169). An Inca prayer recorded by Pachacuti Yamqui states, "I adore you ten times with eyes that blink" ("Relación," p. 320).

88. Murúa, *Origen y genealogía,* p. 257.

89. Sullivan, "Sound and Senses," p. 6.

90. Edmund Leach, *Claude Lévi-Strauss* (New York: Penguin, 1970), pp. 103–104.

91. Hernández Príncipe, "Visitas," pp. 472–473.

92. Isbell writes that such sensory patterning in South American indigenous ritual is used "to create stereotypes and emblems that provide models for the individual's journey" ("Metaphoric Process," p. 305).

93. In this regard it is interesting to note that the Incas apparently made little or limited use of hallucinogens, which characteristically produce a mixing of sensory modalities; see Marlene Dobkin de Ríos, *Hallucinogens: Cross-Cultural Perspectives* (Albuquerque: University of New Mexico Press, 1984), ch. 10, "The Inca of Peru," pp. 149–159. If the Incas did make use of certain hallucinogens, their myths and rituals nonetheless display little of the sensory mingling characteristic of the hallucinogen-using cultures of the Amazon; see Constance Classen, "Sweet Colors, Fragrant Songs: Sensory Models of the Andes and the Amazon," *American Ethnologist* 17:4 (1990):728.

CHAPTER 6 ILLNESS AND DEATH

1. Garcilaso de la Vega discusses Inca medicine in *Comentarios reales,* vol. 1, pp. 115–118.

2. Polo de Ondegardo, "Errores y supersticiones," pp. 12–13.

3. Arriaga, *Extirpation of Idolatry,* p. 48.

4. Cobo, "Nuevo Mundo," p. 234.

5. Polo de Ondegardo, "Errores y supersticiones," pp. 14–15.

6. Ibid., p. 14.

7. Ibid., p. 13.

8. Ibid.

9. For example, the huacas called Sonconancay and Intillapa on the Chinchasuyu quarter of the ceques; see Cobo, "Nuevo Mundo," p. 170.

10. Ibid., pp. 217–218. The participants in this rite also carried burning torches that they threw into streams. Garcilaso de la Vega writes that, "if on a following day an Indian, whatever his age, came across one of these torches in a stream, he fled from it as though from fire, so that the evil that had been banished with it might not be passed on to him" (*Comentarios reales,* vol. 2, p. 100).

11. Cobo, "Nuevo Mundo," pp. 217–218.

12. Ibid., pp. 218–219.

13. Ibid., p. 219.

14. Bastien, *Healers of the Andes,* p. 42.

15. See Irene Silverblatt, "The Evolution of Witchcraft and the Meaning of Healing in Colonial Andean Society," *Culture, Medicine and Psychiatry* 7 (1983):418. This holds true in the modern Andes as well; see, e.g., Bastien, *Mountain of the Condor,* p. 129.

16. Cobo, "Nuevo Mundo," p. 218.

17. Ibid., p. 172.

18. Bastien, *Mountain of the Condor,* pp. 45–46.

19. Guaman Poma, *El primer nueva corónica,* vol. 1, p. 227.

20. George Urioste, "Sickness and Death in PreConquest Andean Cosmology," in *Health in the Andes,* ed. J. Bastien and J. Donahue (Washington, D.C.: American Anthropological Association, 1981), p. 15.

21. Guaman Poma, *El primer nueva corónica,* vol. 1, p. 256.

22. Cobo, "Nuevo Mundo," p. 274.

23. Bastien, *Mountain of the Condor,* p. 173.

24. Arriaga, *Extirpation of Idolatry,* p. 56.

25. Cobo, "Nuevo Mundo, pp. 274–275. In the modern Andes a domestic animal, such as a dog, is sometimes killed and buried with a deceased person to act as his or her companion in the afterlife. See Juan B. Ambrosetti, *Supersticiones y leyendas: región misionera, valles calchaquíes, las pampas* (Buenos Aires: Pingüino, 1947), p. 97; Tschopik, "Aymara of Chucuito," p. 217.

26. Guaman Poma, *El primer nueva corónica,* vol. 1, p. 231.

27. Ibid. See also Arriaga, *Extirpation of Idolatry,* p. 56. For a description of the Feast of the Dead in the modern Andes see Bastien, *Mountain of the Condor:* "Among modern Andeans this feast is held on November 2, the Catholic All Soul's Day, an overlapping of Andean and Catholic ritual that began soon after the Conquest" (p. 178).

28. Arriaga, *Extirpation of Idolatry,* pp. 56–57. A similar distrust of church cemeteries exists among modern Andeans. Tschopik reports an Aymara woman as saying that "the dead should be buried where roads cross. . . . There will be a time when they will all live again, and it is better to bury them by the roads. I don't think that those in the cemeteries will ever live again" ("Aymara of Chucuito," p. 217).

29. Arriaga, *Extirpation of Idolatry,* p. 55; Pachacuti Yamqui, "Relación," pp. 243–244. Tschopik reports of the modern Aymara that "the soul in every way resembles the body and duplicates the actions of the living. Indeed, so great is the resemblance that informant 39 was led to observe: 'Unless you know a man is dead, it is sometimes difficult to say whether you have seen him or whether it was his soul you saw.'" ("Aymara of Chucuito," p. 211).

30. Henrique-Osvaldo Urbano, "Dios Yaya, Dios Churi y Dios Espíritu: mod-

elos trinitarios y arqueología mental en los Andes," *Journal of Latin American Lore* 6:1 (1980):124.

31. See Custred, "Inca Concepts of Soul and Spirit."

32. Titu Cusi Yupanqui, *Ynstrución,* p. 31.

33. Guaman Poma, *El primer nueva corónica,* vol. 1, p. 227.

34. Jorge A. Lira, *Medicina andina: farmacopea y ritual* (Cuzco: Centro de Estudios Rurales Andinos Bartolomé de Las Casas, 1985), p. 86.

35. Arriaga, *Extirpation of Idolatry,* p. 55.

36. Ambrosetti, *Supersticiones y leyendas,* p. 97.

37. Mishkin, "Contemporary Quechua." For data on this from the Bolivian Andes see Oblitas Poblete, *Cultura Callawaya,* pp. 32–33; Bastien, *Healers,* p. 72.

38. Catherine Allen, "Body and Soul in Quechua Thought," *Journal of Latin American Lore* 8:2 (1982):184–187. Mishkin writes that the Andeans of Kauri, Peru, believe that a body is composed of wet flesh, a soul that resides in the bones, and a spirit that goes to heaven after death ("Contemporary Quechua," p. 465).

39. Arriaga mentions that he asked a "soul-eating" Andean sorcerer what a soul tasted like and was told that it was like "dried beef" (*Extirpation of Idolatry,* p. 39). This seems to associate the soul with dryness; however, it is impossible to tell exactly what Andean entity corresponded to the Catholic concept of soul.

40. Conrad and Demarest, *Religion and Empire,* p. 105.

41. Bastien, *Mountain of the Condor,* p. 172.

42. Juan Polo de Ondegardo, "Relación de los fundamentos acerca del notable daño que resulta de no guardar a los indios sus fueros" [1571], in *Colección de libros y documentos referentes a la historia del Perú,* ser. 1, vol. 3, ed. H. H. Urteaga and C. A. Romero (Lima: Sanmartí, 1916), p. 111.

43. The anonymous chronicler gives, as the primary reason for the Inca custom of preserving the bodies of the dead, the Inca belief that "the spirits would return to their bodies after a certain time and resuscitate them, and this could not occur unless the bodies were kept uncorrupted, with nothing missing, at least of their bones, since the flesh might disintegrate" ("Costumbres antiguas," pp. 159–160). See also Garcilaso de la Vega, *Comentarios reales,* vol. 1, p. 80.

44. Garcilaso de la Vega, *Comentarios reales,* vol. 1, pp. 286–287. The Inca mummies appropriated by the Spanish were taken to the Hospital de Españoles de San Andrés in Lima where they were exhibited for some time and then buried somewhere on the grounds of the hospital; see José de la Riva-Agüero, *Obras completas de José de la Riva-Agüero,* vol. 5 (Lima: Pontificia Universidad Católica del Perú, 1966), p. 397.

45. Mummification was already a time-honored and widespread practice in the Andes by the rise of the Incas. See Aidan Cockburn and James M. Vreeland, Jr., "Mummies of Peru," in *Mummies, Disease and Ancient Cultures,* ed. A. and E. Cockburn (Cambridge: Cambridge University Press, 1980). As they did with so many other Andean customs, the Incas developed this practice into one of particular significance in their imperial and cosmological order.

46. Another benefit of mummification is that the mummies would not smell rotten (Cobo, "Nuevo Mundo," p. 163), as ordinary corpses would, an important consideration in a society in which the smell of rottenness was equated with chaos and savagery.

47. Ibid., p. 162. Pierre Duviols discusses the role of the huaque in the Andes in "Un symbolisme andin du double, la lithomorphose de l'ancêtre."

48. Cobo, "Nuevo Mundo," p. 67.
49. Sarmiento de Gamboa, "Historia Índica," p. 236.
50. Garcilaso de la Vega, *Comentarios reales,* vol. 1, pp. 173–174.
51. Cobo names the place where each Inca mummy was found at the end of his account of that Inca's life; see "Nuevo Mundo," pp. 64–94.
52. Ibid., p. 164.
53. Ibid., p. 66.
54. Pizarro writes of the caretakers of the Inca mummies that "whenever they wished to eat, to drink, they said that the dead ones wished to do that same thing. If they wished to go and divert themselves in the houses of other dead folk, they said the same, for it was customary for the dead to visit one another" (*Discovery and Conquest,* vol. 1, p. 203).
55. Pachacuti Yamqui, who relates this account, adds that to punish his sister for not marrying him Huayna Capac gave her to an ugly, old, coca-addicted curaca (chieftain). However, she chose instead to become an aclla, indicating that Inca women had at least some power of self-determination. See "Relación," pp. 259–260.
56. Murúa, *Historia general,* p. 122. Huascar is also said to have married his mother to his father's mummy in order to ensure his own legitimacy; see Pachacuti Yamqui, "Relación," p. 266.
57. Cobo writes that the ashes of the hearts of the dead Incas were kept in the abdomen of a statue of the Sun in Coricancha ("Nuevo Mundo," p. 157). As the heart stood for the person in Andean thought, this would effectively associate the dead Incas with the Sun.
58. Sarmiento de Gamboa, "Historia Índica," pp. 223–224.
59. Cobo, "Nuevo Mundo," p. 73. In the present-day Cuzco region leaving a corpse exposed on the ground is supposed to prevent rain from falling; see Lira, *Medicina andina,* p. 85; Tschopik, "Aymara of Chucuito," p. 214. The corpse here is perhaps thought to "scare" the rain away; however, there is also the possibility that the presence of the dead in the world of the living is thought to create a certain sterility (lifelessness), exemplified by drought, in the latter; see Tschopik, "Aymara of Chucuito," p. 214.
60. Cieza, *El señorío de los Incas,* p. 104.
61. Cobo, "Nuevo Mundo," p. 164.
62. Pizarro, *Discovery and Conquest,* vol. 1, pp. 205–206.
63. Conrad and Demarest, *Religion and Empire,* p. 126.

CHAPTER 7 THE METAPHORICAL BODY

1. Garcilaso de la Vega, *Comentarios reales,* vol. 1, p. 43.
2. That a relationship existed between the number of panacas in each half of Cuzco and the number of fingers on a hand is not explicitly stated in the chronicles, but seems a likely reason for the arbitrary division of each moiety into five parts. Zuidema writes of the prevalence of quinque-partition among the Incas that "it can only be assumed that the decimal system of counting in the Peruvian languages, and the number of fingers on a hand, influenced this use of the number five" (*The Ceque System of Cuzco,* p. 214).
3. Interestingly, the Incas seem to have associated the lower, or southern, half of their empire with fluidity (Lake Titicaca, the mythological birthplace of humanity, was situated there, for example) and the upper, or northern, half

with structure (Huayna Capac built his new capital city in Quito). Such associations would be consistent with those attached to the upper and lower halves of the human body.

4. Zuidema, "Masks," p. 151.

5. Betanzos, "Suma y narración," p. 50.

6. Murúa, *Origen y genealogía,* p. 164. Other chroniclers (e.g., Guaman Poma, *El primer nueva corónica,* vol. 1, p. 304; Garcilaso de la Vega, *Comentarios reales,* vol. 2, p. 13) write that persons from a particular locality were trained to carry the Inca's litter. The practice described by Murúa was therefore likely one of ritual, rather than everyday, use.

7. Murúa, *Origen y genealogía,* p. 164.

8. Garcilaso de la Vega, *Comentarios reales,* vol. 2, p. 13.

9. Ibid., vol. 2, p. 106.

10. Fernando de Montesinos, "Memoriales antiguas historiales y políticas del Perú" [1644], in *Colección de libros y documentos referentes a la historia del Perú,* ser. 2A, vols. 4–7, ed. H. Urteaga (Lima: Sanmartí, 1911), p. 37. Montesinos names Incas not referred to by other chroniclers.

11. Ibid., p. 77.

12. Guaman Poma, *El primer nueva corónica,* vol. 3, p. 913.

13. Garcilaso de la Vega, *Comentarios reales,* vol. 1, p. 89. Modern Andeans also think of Cuzco as a navel; see Abraham Valencia Espinoza, "Inka Qollari dramatizado," in *Ideología mesiánica del mundo andino,* ed. J. Ossio A. (Lima: Colección Biblioteca de Antropología, 1973), p. 291. The exact etymology of the word *cuzco,* however, is uncertain; see Raúl Porras Barrenechea, "Prólogo," in *Antología del Cuzco,* ed. R. Porras Barrenechea (Lima: Librería Internacional del Perú, 1961), pp. xvii–xviii.

14. Duviols, "La capacocha," p. 24.

15. Betanzos, "Suma y narración," p. 40.

16. Garcilaso de la Vega, *Comentarios reales,* vol. 2, pp. 7–12; Murúa, *Historia general,* p. 124.

17. Cieza, *El señorío de los Incas,* pp. 172–173.

18. Montesinos, "Memoriales," p. 107.

19. Betanzos, "Suma y narración," p. 50.

20. Ibid., p. 48.

21. Sarmiento de Gamboa, "Historia Índica," p. 258.

22. R. T. Zuidema, "The Lion in the City: Royal Symbols of Transition in Cuzco," *Journal of Latin American Lore* 9:1 (1983):89.

23. Pachacuti Yamqui, "Relación," p. 240.

24. Cieza, *El señorío de los Incas,* p. 139. Guaman Poma writes that Pachacuti had "lion's eyes" (*El primer nueva corónica,* vol. 1, p. 89).

25. Molina of Cuzco, "Fábulas y ritos," pp. 20–21.

26. Ibid., p. 58.

27. Zuidema, "The Lion in the City," p. 40.

28. Cobo, "Nuevo Mundo," p. 177.

29. Garcilaso de la Vega, *Comentarios reales,* vol. 1, p. 105.

30. In the modern Peruvian community of Sonqo the Incas are believed to live in a city guarded by lions in the jungle; see Wagner, "Coca, Chicha and Trago," p. 80.

31. Guaman Poma, *El primer nueva corónica,* vol. 2, p. 655. For an analysis of the opposition between pumas and jaguars in Inca culture, see Zuidema, "The Lion in the City," pp. 81–88.

32. Ralph Bolton, "Nine Quechua-Qolla Narratives: Peruvian Folktale Texts," *Anthropos* 75 (1980):150–151.

33. Cobo, "Nuevo Mundo," p. 172.

34. Urioste, *Hijos de Pariya Qaqa,* vol. 2, p. 185.

35. Urton, "Animal Metaphors," p. 255.

36. John H. Rowe, "What Kind of Settlement Was Inca Cuzco?" *Ñawpa Pacha* 5 (1967).

37. Santiago Agurto Calco, *Cusco: la traza urbana de la ciudad inca* (Cuzco: UNESCO, Instituto Nacional de Cultura del Perú, Proyecto Per-71/539, 1980), pp. 112, 138.

38. Zuidema, "The Lion in the City," p. 68.

39. Tschopik, "Aymara of Chucuito," p. 152.

40. Javier Albo et al., "Dinámica en la estructura inter-comunitaria de Jesús de Machaca," *América Indígena* 32:3 (1972):788–790.

41. For an exploration of body metaphors, see Mark Johnson, *The Body in the Mind: The Bodily Basis for Meaning, Imagination, and Reason* (Chicago: University of Chicago Press, 1987).

42. Antonio de la Calancha, *Crónica moralizado del orden de San Agustín en el Perú con sucesos ejemplares en esta monarquía* [1638], Vol. 3, ed. I. Prado Pastor (Lima: Universidad Nacional Mayor de San Marcos, 1974–1981), p. 932.

43. Bastien, *Mountain of the Condor,* p. xix.

44. Ibid., p. 29.

45. Bastien explores this metaphor in his *Healers,* also.

46. Ortiz Rescaniere, "El mito de la escuela," p. 239. "Lima" is a corruption of the Quechua word *rimac,* which means "speaker."

47. Condori and Gow, *Kay Pacha,* pp. 10–12.

48. Guaman Poma, *El primer nueva corónica,* vol. 1, p. 249.

49. Murúa, *Origen y genealogía,* p. 305.

50. Stark, "Quechua Body Parts," pp. 8–9.

51. Polo de Ondegardo, "Errores y supersticiones," pp. 3–5.

52. Pizarro writes that more than two hundred women slept in the temple of the Sun, "and they pretended . . . that the Sun had connexion with them" (*Discovery and Conquest,* vol. 1, pp. 255–256).

53. William H. Isbell has made a similar analysis of the buildings surrounding the Aucaypata square of Cuzco. He concludes that the houses of learning on the right side of the square were associated with masculinity, the house of chosen women and the palace called Amarucancha (Enclosure of Serpents) on the left side were associated with femininity, and the buildings dedicated to Viracocha at the apex with bisexuality. (There were no buildings on the south side of the plaza.) See "Cosmological Order Expressed in Prehistoric Ceremonial Centers," *Actes du XLIIe Congrès International des Américanistes,* vol. 4 (Paris, 1976), p. 275.

This analysis can also be applied to the three institutions discussed here: The temple of the Sun is male; the house of the chosen women is female; and the Inca is a submissive female in relation to the former and a dominant male in relation to the latter.

54. Garcilaso de la Vega, *Comentarios reales,* vol. 1, p. 66.

55. Bastien, *Mountain of the Condor,* p. 47.

56. Urioste, *Hijos de Pariya Qaqa,* vol. 2, p. 195.

57. Bastien, *Mountain of the Condor,* p. 86.

58. Still another association could be made of the outside of the body with

structure, clarity, and the past, and the inside of the body with fluidity, darkness, and the future. This would explain why the Incas would examine the entrails of animals in order to learn of the future and the unknown. See Polo de Ondegardo, "Instrución," pp. 196–197.

59. Urioste, *Hijos de Pariya Qaqa,* vol. 2, p. 177.

CHAPTER 8 THE DISORDERED BODY

1. Cieza, *El señorío de los Incas,* p. 165.

2. Garcilaso de la Vega, *Comentarios reales,* vol. 2, p. 250.

3. The hierarchically oriented Andeans noted in amazement that the Spanish "all seemed like brothers in their way of dressing and speaking with each other"; Guaman Poma, *El primer nueva corónica,* vol. 2, p. 354.

4. Molina el Almagrista, "Relación," p. 87.

5. Titu Cusi Yupanqui, *Ynstrución,* p. 17.

6. A comprehensive history of the Incas in the post-Conquest period is provided by Philip A. Means in *Fall of the Inca Empire and the Spanish Rule in Peru 1530–1780* (New York: Gordian Press, 1964).

7. Wachtel, *Vision of the Vanquished,* pp. 86–90.

8. Arriaga, *Extirpation of Idolatry,* p. 19.

9. Ibid., p. 25. Arriaga adds that the Indians must not know the whereabouts of the remains of their huacas: "It is well known, for instance, that the Indians of Huaylas, although a long way away, worshipped on a bridge in Lima because some huacas that Francisco Cano had taken away from them had been thrown into the river there" (*Extirpation of Idolatry,* p. 26).

10. Cobo, "Nuevo Mundo," p. 77.

11. Ibid.

12. Garcilaso de la Vega, *Historia general del Perú* [1617], vol. 3, ed. A. Rosenblat (Buenos Aires: Emecé, 1944), p. 246.

13. This was the most important of a series of native revolts in the eighteenth and early nineteenth centuries in the Andes.

14. Lilian Fisher discusses this revolt and its consequences for the Andeans in *The Last Inca Revolt: 1780–1783* (Norman: University of Oklahoma Press, 1966).

15. Guaman Poma, *El primer nueva corónica,* vol. 2, p. 354.

16. According to Guaman Poma, this miscegenation was the worst threat to Andean culture. See his *El primer nueva corónica,* vol. 3, pp. 909–910.

17. The concept of an upside-down world in which everything functioned in reverse to the ordinary world was a commonplace in Europe of the time. See, e.g., David Kunzle, "World Upside Down: The Iconography of a European Broadsheet Type," in *The Reversible World: Symbolic Inversion in Art and Society,* ed. Barbara Babcock (Ithaca, N.Y.: Cornell University Press, 1978). Guaman Poma's use of this concept does not necessarily reflect a European influence, however, as the notion of a world reversal was fundamental to Andean cosmology.

18. See, e.g., Guaman Poma, *El primer nueva corónica,* vol. 2, pp. 502, 711.

19. Ibid., vol. 2, pp. 820–821.

20. Murúa, *Historia general,* p. 78.

21. Guaman Poma, *El primer nueva corónica,* vol. 3, p. 879.

22. Pierre Duviols, ed., *Cultura andina y represión: procesos y visitas de idolatrías y hechicerías, Cajatambo, siglo XVII* (Cusco: Centro de Estudios Rurales Andinos

Bartolomé de las Casas, 1986), pp. 257, 314. For a depiction of this punishment, see Guaman Poma, *El primer nueva corónica,* vol. 2, p. 488.

23. Murúa. *Origen y genealogía,* p. 353.

24. Guaman Poma, *El primer nueva corónica,* vol. 2, p. 377.

25. Ibid., vol. 1, p. 74.

26. Guaman Poma describes Viracocha's namesake, Viracocha Inca, as white and bearded (*El primer nueva corónica,* vol. 1, p. 87). The Spanish, as the "sons of Viracocha" (Titu Cusi Yupanqui, *Ynstrución,* p. 17), may well have been seen by the Incas as reversing the overthrow of Viracocha occasioned by Inca Pachacuti, the son of the Sun.

27. Garcilaso de la Vega, *Comentarios reales,* vol. 1, p. 86, and *Historia general del Perú,* vol. 1, p. 42.

28. The conception of the Spanish as "wild" was originally established in the minds of the Andeans by the very appearance of the conquistadores, dressed as they were in armor and mounted on terrifying beasts.

29. Urioste, *Hijos de Pariya Qaqa,* vol. 1, pp. 19, 37. The corpses of criminals, who embodied anticulture, were left to be eaten by wild animals in the Inca empire (see, e.g., Guaman Poma, *El primer nueva corónica,* vol. 1, p. 163).

30. Allen, "Patterned Time," p. 156.

31. Titu Cusi Yupanqui, *Ynstrución,* pp. 17–18.

32. Guaman Poma, *El primer nueva corónica,* vol. 2, p.347. Columbus, in fact, was so obsessed with discovering gold that he interpreted almost every characteristic—the beauty of the land, the warmth of the climate, etc.—of the New World as a sign of its presence; see Todorov, *Conquest of America,* pp. 20–21.

33. Molina el Almagrista writes that the Incas thought gold was the tears of the sun. He also mentions that when a particularly large nugget of gold was found it was placed in a shrine, for it was believed that it would attract other gold to it ("Relación," p. 76).

34. Garcilaso de la Vega, *Comentarios reales,* vol. 2, pp. 7–12.

35. There is an abundant literature on the effects of literacy on consciousness; e.g., Walter Ong, *Orality and Literacy: The Technologizing of the Word* (London: Methuen, 1982). My intention here is not to enter that debate, but simply to look at the Andean perception of writing and to situate this within the larger context of the clash between sensory orders. See also Constance Classen, "Literacy as Anti-Culture: The Andean Experience of the Written Word," *History of Religions* 30:4 (1991).

36. Lewis Hanke, *The Spanish Struggle for Justice in the Conquest of America* (Philadelphia: University of Pennsylvania Press, 1949), p. 6.

37. The Incas apparently also kept pictorial records. See Polo de Ondegardo, "Relación," p. 57; and Garcilaso de la Vega, *Comentarios reales,* vol. 1, p. 84.

38. Marcia and Robert Ascher, *Code of the Quipu: A Study in Media, Mathematics and Culture* (Ann Arbor: University of Michigan Press, 1981), p. 62.

39. E.g., Marshall McLuhan, *Understanding Media: The Extension of Man* (New York: McGraw-Hill, 1964); Walter Ong, *The Presence of the Word* (New Haven, Conn.: Yale University Press, 1967).

40. Titu Cusi Yupanqui, *Ynstrución,* p. 2. The Spanish likely often read aloud to themselves, rather than silently.

41. Guaman Poma, *El primer nueva corónica,* vol. 2, p. 353.

42. Titu Cusi Yupanqui, *Ynstrución,* p. 4.

43. Murúa, *Historia general,* p. 176.

44. Garcilaso de la Vega, *Historia general del Perú,* vol. 1, pp. 98–99.
45. Montesinos, "Memoriales," p. 67.
46. Arriaga, *Extirpation of Idolatry,* p. 126.
47. Ibid., pp. 123–124.
48. Claude Lévi-Strauss, *Tristes Tropiques,* trans. J. Russell (New York: Atheneum, 1967), p. 292.
49. Arriaga, *Extirpation of Idolatry,* p. 66.
50. Guaman Poma's use of writing and illustrations to denounce Spanish oppression is explored by Rolena Adorno in *Guaman Poma: Writing and Resistance in Colonial Peru* (Austin: University of Texas Press, 1986).
51. Guaman Poma, *El primer nueva corónica,* vol. 3, p. 889.
52. The failure of the Spanish to keep their word is stressed in Titu Cusi Yupanqui's memorial. He relates that his father, Manco Inca, said to him on his deathbed, "Don't allow [the Spanish] to enter your land, however much they ply you with words, for their honeyed words deceived me and they will do the same to you if you believe them" (*Ynstrución,* p. 31).
53. Guaman Poma, *El primer nueva corónica,* vol. 2, p. 656 (emphasis added). The Andeans soon became caught up in the fetishistic Spanish legal system, expending their time and money in lengthy lawsuits, often over the ownership of land. George Kubler writes that "litigation exercised a fascination over the Indians; it was regarded as a talisman, and to the possessor it made little difference whether the paper resolved the litigation for or against his interests." ("The Quechua in the Colonial World," in *Handbook of South American Indians,* vol. 2: *The Andean Civilizations,* ed. J. Steward [New York: Cooper Square, 1963], p. 379). Tschopik reports that among the modern Andeans of Chucuito, Bolivia, litigation has largely replaced sorcery as a way of seeking retribution ("Aymara of Chucuito," pp. 170, 181).
54. Guaman Poma, *El primer nueva corónica,* vol. 2, p. 729.
55. Polo de Ondegardo, "Instrución," p. 202.
56. Titu Cusi Yupanqui, *Ynstrución,* p. 26.
57. Garcilaso de la Vega, *Historia general del Perú,* vol. 1, p. 83.
58. Ibid., vol. 1, pp. 83–84.
59. Molina of Cuzco attributes the "blindness" of Andean beliefs in great part to the Andeans' lack of writing ("Fábulas y ritos," p. 17). Guaman Poma describes his ancestors as being "completely mistaken and blind" because they did not know how to read and write (*El primer nueva corónica,* vol. 1, p. 49).
60. Urioste, *Hijos de Pariya Qaqa,* vol. 2, pp. 161–172.
61. Pedro de Cieza de León, *La crónica del Perú,* ed. M. Ballesteros (Madrid: Historia 16, 1984), pp. 393–396.
62. Ibid., p. 396.
63. Guaman Poma, *El primer nueva corónica,* vol. 2, p. 360.
64. Ibid., p. 781.
65. Ibid., p. 804.
66. Molina of Cuzco, "Fábulas y ritos," p. 79.
67. Ibid., pp. 80–81.
68. Guaman Poma, *El primer nueva corónica,* vol. 1, p. 256.
69. Molina of Cuzco, "Fábulas y ritos," p. 80.
70. General Archive of the Indies, Audiencia Lima, legajo 316, 1571 notebook, fos. 32r–33r, evidence of Gerónimo Martín, cited by Wachtel, *Vision of the Vanquished,* p. 267 n. 76.
71. Ibid., p. 181. The Taqui Oncoy movement has much in common with

other millenarian movements around the world. See, e.g., Sylvia Thrupp, ed., *Millennial Dreams in Action: Studies in Revolutionary Religious Movements* (New York: Schocken Books, 1970).

72. Molina of Cuzco, "Fábulas y ritos," p. 82.

73. Garcilaso de la Vega, *Historia general del Perú,* vol. 3, pp. 249–250. See also Murúa, *Historia general,* p. 271; and Baltasar de Ocampo, "Descripción y sucesos históricos de la provincia de Vilcabamba," in *Colección de libros y documentos referentes a la historia del Perú,* ser. 2A, vols. 4–7 (Lima: Sanmartí, 1911), pp. 172–173.

74. Ocampo, "Vilcabamba," p. 175.

CHAPTER 9 CONCLUSION

1. Garcilaso de la Vega, *Comentarios reales,* vol. 1, p. 265.

2. Ibid., vol. 1, p. 174.

3. This is also the order, minus touch, that Guaman Poma follows in his prayer for a Christian sensory order (see Chapter 8).

4. It might even be possible to interpret the Inca diagram of the cosmos (see figs. 1 and 2) as a face if one sees the sun and moon as eyes, Viracocha as a nose (breathing life into the cosmos?), and the grid of cultivated earth at the bottom as a mouth. This "face" actually resembles the face of the sun in the diagram.

5. See González Holguín, *Vocabulario, ppichiu,* p. 284.

6. In fact, the Incas are said to have received omens forecasting the fall of their empire; see Garcilaso de la Vega, *Comentarios reales,* vol. 2, p. 249.

7. See Classen, "Literacy as Anti-Culture."

8. "Tragedia del fin de Atahualpa," quoted by Wachtel, *Vision of the Vanquished,* p. 39.

9. Ibid., p. 40.

10. Wagner, "Coca, Chicha and Trago," p. 47.

11. Condori and Gow, *Kay Pacha,* p. 56.

12. Ortiz Rescaniere, "El mito de la escuela."

13. Polo de Ondegardo lists as a common Andean "error against the Catholic faith" the belief that "Jesus Christ our Lord and the Devil [the Andean deities?] can both be worshipped because they have now reconciled and become as brothers" ("Instrución," p. 202).

14. Ortiz Rescaniere, "El mito de la escuela," p. 241.

15. See, e.g., Tschopik, "Aymara of Chucuito," pp. 190ff., 197, 206ff. God and the sky are usually thought of as too remote to take much interest in human affairs. In the words of one Andean, "[Father God] sends the rain. He doesn't care. He's sitting up there on top of the clouds, nice and warm, while we're down here freezing" (Wagner, "Coca, Chicha and Trago," p. 58).

16. Bastien, *Mountain of the Condor,* p. 97.

17. Ibid., p. 96; Mishkin, "Contemporary Quechua," p. 464.

18. Urbano, "Dios Yaya, Dios Churi y Dios Espíritu," p. 121. In this regard, Nash writes that the Andean miners of Bolivia postulate a dichotomy between the post-Conquest Christian deities and the pre-Conquest Andean ones, with the former operating aboveground and the latter below (*We Eat the Mines and the Mines Eat Us,* p. 7).

19. Wagner, "Coca, Chicha and Trago," pp. 80–81.

20. Urbano, "Dios Yaya," pp. 111–112. This tripartite division of time is

reminiscent of that made by the twelfth-century Calabrian abbot, Joachim of Flora. He also divided history into three ages, one of the Father, one of the Son, and one of the Holy Spirit (which he believed would begin in the year 1260).

21. To determine the extent of this modification, it would be necessary to know whether the Andeans consider this third age to be the end of time, or whether other ages are thought possibly to follow it.

22. Many of these legends are collected in J. Ossio A., ed., *Ideología mesiánica del mundo andino* (Lima: Colección Biblioteca de Antropología, 1973).

23. David Gow, "The Roles of Christ and Inkarri in Andean Religion," *Journal of Latin American Lore* 6:2 (1980); Rosalind Gow, "Inkarri and Revolutionary Leadership in the Southern Andes," *Journal of Latin American Lore* 8:2 (1982).

24. R. Gow, "Inkarri and Revolutionary Leadership," p. 212.

25. Ibid., p. 211.

26. José María Arguedas, *Formación de una cultura nacional indoamericana* (México: Siglo Veintiuno, 1975), p. 40.

27. Ibid., p. 41.

28. Ibid., p. 178.

29. The observation that social conditioning extends to some of our most basic physiological functions, such as sleeping and eating, was made by Marcel Mauss in his landmark essay "Body Techniques," in *Sociology and Psychology Essays*, trans. B. Brewster (London: Routledge and Kegan Paul, 1979).

30. Douglas, *Natural Symbols*.

31. Michael Jackson, "Thinking Through the Body: An Essay on Understanding Metaphor," *Social Analysis* 14 (December 1983):143.

32. Ibid., p. 137.

33. For a guide on how to study sensory symbolism, see Constance Classen and David Howes, "Conclusion: Sounding Sensory Profiles," in *The Varieties of Sensory Experience: A Source Book in the Anthropology of the Senses*, ed. David Howes (Toronto: University of Toronto Press, 1991).

34. For a comparison of the sensory models of two oral cultures, see Classen, "Sweet Colors, Fragrant Songs."

35. Ong, *Presence of the Word*, pp. 117–118; Donald Lowe, *History of Bourgeois Perception* (Chicago: University of Chicago Press, 1982), pp. 5–7.

36. Sam Gill, *Mother Earth: An American Story* (Chicago: University of Chicago Press, 1987), p. 163.

APPENDIX QUECHUA BODY AND SENSORY TERMS

1. Ángel Avedaño, *La rebellion de los mallkis: medicina popular quechua* (Lima: Antwara, 1988). The modern Quechua anatomical terminology provided by Avedaño is more extensive than that given here.

2. For examples of modern Quechua onomatopoeia see Nuckolls, "Sound Symbolic Discourse as Cinematic Text."

BIBLIOGRAPHY

Adorno, Rolena. *Guaman Poma: Writing and Resistance in Colonial Perú.* Austin: University of Texas Press, 1986.

———, ed. *From Oral to Written Expression: Native Andean Chronicles of the Early Colonial Period.* Syracuse, N.Y.: Syracuse University Maxwell School of Citizenship and Public Affairs, 1982.

Agurto Calco, Santiago. *Cusco: la traza urbana de la ciudad inca.* Cuzco: UNESCO, Instituto Nacional de Cultura del Perú, Proyecto Per–71/539, 1980.

Albo, Javier, et al. "Dinámica en la estructura inter-comunitaria de Jesús de Machaca." *América Indígena* 32:3 (1972):773–816.

Allen, Catherine. "Body and Soul in Quechua Thought." In *Journal of Latin American Lore* 8:2 (1982):179–196.

______. "Patterned Time: The Mythic History of a Peruvian Community." In *Journal of Latin American Lore* 10:2 (1984):151–173.

Ambrosetti, Juan B. *Supersticiones y leyendas: región misionera, valles calchaquíes, las pampas.* Buenos Aires: Pingüino, 1947.

Anonymous. "Relación de las costumbres antiguas de los naturales del Pirú" [ca. 1600]. In *Crónicas peruanas de interés indígena,* pp. 152–189. Ed. F. Esteve Barba. Biblioteca de Autores Españoles, vol. 209. Madrid: Ediciones Atlas, 1968.

Arguedas, José María. *Formación de una cultura nacional indoamericana.* México: Siglo Veintiuno, 1975.

______. "Taki Parwa y la poesía quechua de la república." *Letras Peruanas* 12:8 (1955):73–75.

______. *Yawar Fiesta.* Trans. F. Horning Barraclough. Austin: University of Texas Press, 1985.

Arriaga, Pablo Joseph [José]de. *The Extirpation of Idolatry in Peru* [1621]. Trans. and ed. L. Clark Keating. Lexington: University of Kentucky Press, 1968.

Ascher, Marcia, and Robert Ascher. *Code of the Quipu: A Study in Media, Mathematics and Culture.* Ann Arbor: University of Michigan Press, 1981.

Avedaño, Ángel. *La rebellión de los mallkis: medicina popular quechua.* Lima: Antwara, 1988.

Barkan, Leonard. *Nature's Work of Art: The Human Body as Image of the World.* New Haven, Conn.: Yale University Press, 1975.

Bastien, Joseph. *Healers of the Andes.* Salt Lake City: University of Utah Press, 1987.

________. *Mountain of the Condor: Metaphor and Ritual in an Andean Ayllu.* St. Paul, Minn.: West, 1978.

Bennet, Wendell C. "The Andean Highlands." In *Handbook of South American Indians,* vol. 2: *The Andean Civilizations,* pp. 1–60. Ed. J. H. Steward. Bureau of American Ethnology Bulletin 143. New York: Cooper Square, 1963.

Berthelot, J. M. and M. Drulhe, eds. "Les sociologies et le corps." *Current Sociology* 33:2 (Summer, 1985):1–200.

Bertonio, Ludovico. *Vocabulario de la lengua aymara* [1612]. Cochabamba, Bolivia: Ediciones Ceres, 1984.

Betanzos, Juan de. "Suma y narración de los Incas" [1551]. In *Crónicas peruanas de interés indígena*, pp. 5–56. Ed. F. Esteve Barba. Biblioteca de Autores Españoles, vol. 209. Madrid: Ediciones Atlas, 1968.

Bolton, Charlene, and Ralph Bolton. "Rites of Retribution and Restoration in Canchis." *Journal of Latin American Lore* 2:1 (1976):97–114.

Bolton, Ralph. "Nine Quechua-Qolla Narratives: Peruvian Folktale Texts." *Anthropos* 75 (1980):140–162.

Bouysse-Cassagne, Thérèse. "Urco and Uma: Aymara Concepts of Space." In *Anthropological History of Andean Polities*, pp. 201–227. Ed. J. Murra, N. Wachtel, and J. Revel. Cambridge: Cambridge University Press, 1986.

Brown, C. Mackenzie. "Purana as Scripture: From Sound to Image of the Holy Word in Hindu Tradition." *History of Religions* 26:1 (August 1988):68–86.

Calancha, Antonio de la. *Crónica moralizado del orden de San Agustín en el Perú con sucesos ejemplares en esta monarquía* [1638]. 6 vols. Ed. I. Prado Pastor. Lima: Universidad Nacional Mayor de San Marcos, 1974–1981.

Capoche, Luis. "Relación general de la villa imperial de Potosí" [1585]. In *Relaciones histórico-literarias de la América meridional*, pp. 9–189. Ed. L. Hanke. Biblioteca de Autores Españoles, vol. 122. Madrid: Ediciones Atlas, 1959.

Cassirer, Ernst. *The Philosophy of Symbolic Forms*, vol. 1: *Language*; vol. 2: *Mythical Thought*. Trans. R. Mannheim. New Haven Conn.: Yale University Press, 1953, 1955.

Cereceda, Veronica. "Aproximaciones a una estética andina: de la belleza al tinku." In *Tres reflexiones sobre el pensamiento andino*, pp. 133–231. Ed. J. Medina. La Paz: Hisbol, 1987.

________. "The Semiology of Andean Textiles: The Talegas of Isluga." In *Anthropological History of Andean Polities*, pp. 149–173. Ed. J. Murra, N. Wachtel, and J. Revel. Cambridge: Cambridge University Press, 1986.

Chidester, David Shane. "Word and Light: Perception and Symbolic Forms in the Augustinian Tradition." Ph.D. diss., University of California, Santa Barbara, 1981.

Cieza de León, Pedro de. *La crónica del Perú* [1550]. Ed. M. Ballesteros. Madrid: Historia 16, 1984.

______. *El señorío de los Incas* [1550]. Ed. M. Ballesteros. Madrid: Historia 16, 1985.

Classen, Constance. "Aesthetics and Asceticism in Inca Religion." *Anthropologica* 32:1 (1990):101–106.

________. "Creation by Light/Creation by Sound: A Sensory Analysis of Two South American Cosmologies." In *The Varieties of Sensory Experience: A Source Book in the Anthropology of the Senses*, pp. 239–255. Ed. D. Howes. Toronto: University of Toronto Press, 1991.

________. "Literacy as Anti–Culture: The Andean Experience of the Written Word." *History of Religions* 30:4 (1991): 404–421.

________. "Sweet Colors, Fragrant Songs: Sensory Models of the Andes and the Amazon." *American Ethnologist* 17:4 (1990):722–735.

________, and Howes, David. "Conclusion: Sounding Sensory Profiles." In *The Varieties of Sensory Experience: A Source Book in the Anthropology of the Senses*, pp. 257–288. Ed. David Howes. Toronto: University of Toronto Press, 1991.

Cobo, Bernabé. "Historia del Nuevo Mundo" [1653]. In *Obras del P. Bernabé Cobo*, vol. 2, pp. 7–275. Ed. F. Mateos. Biblioteca de Autores Españoles, vol. 92. Madrid: Ediciones Atlas, 1964.

Cockburn, Aidan, and James M. Vreeland, Jr. "Mummies of Perú." In *Mummies, Disease and Ancient Cultures*, pp. 135–174. Ed. A. and E. Cockburn. Cambridge: Cambridge University Press, 1980.

Condori, Bernabé, and Rosalind Gow. *Kay Pacha.* Cusco, Perú: Centro de Estudios Rurales Andinos Bartolomé de Las Casas, 1982.

Conrad, Geoffrey, and Arthur Demarest. *Religion and Empire: The Dynamics of Aztec and Inca Expansionism.* Cambridge: Cambridge University Press, 1984.

Corbin, Alain. *The Foul and the Fragrant: Odour and the French Social Imagination.* Trans. M. Kochan, R. Porter, and C. Prendergast. Cambridge, Mass.: Harvard University Press, 1988.

Custred, Glynn. "Inca Concepts of Soul and Spirit." In *Essays in Humanistic Anthropology: A Festschrift in Honor of David Birney*, pp. 277–302. Ed. B. T. Grindal and D. M. Warren. Washington, D.C.: University Press of America, 1979.

Dobkin de Ríos, Marlene. *Hallucinogens: Cross-Cultural Perspectives.* Albuquerque: University of New Mexico Press, 1984.

Douglas, Mary. *Natural Symbols: Explorations in Cosmology.* New York: Pantheon, 1970.

Duviols, Pierre. "La capacocha." *Allpanchis* 9 (1976):11–54.

______. *La lutte contre les religions autochtones dans le Pérou colonial; "l'extirpátion de l'idolâtrie" entre 1532 et 1660.* Lima: Institut Français d'Études Andines, 1972.

______. "Un symbolisme andin du double, la lithomorphose de l'ancêtre." In *Actes du XLIIe Congrès Internacional des Américanistes*, pp. 359–364. Paris, 1976.

_______, ed. *Cultura andina y represión: procesos y visitas de idolatrías y hechicerías, Cajatambo, siglo XVII* [1600s]. Cusco: Centro de Estudios Rurales Andinos Bartolomé de Las Casas, 1986.

Earls, John, and Irene Silverblatt. "La realidad física y social en la cosmología andina." In *Actes du XLIIe Congrès International des Américanistes*, vol. 4, pp. 299–325. Paris, 1976.

Eck, Diana. *Darsan: Seeing the Divine Image in India.* Rev. 2d ed. Chambersburg, Penn.: Anima Books, 1985.

Fisher, Lillian. *The Last Inca Revolt: 1780–1783.* Norman: University of Oklahoma Press, 1966.

Garcilaso de la Vega, "El Inca." *Comentarios reales de los Incas* [1609]. 2 vols. Ed. A. Rosenblat. Buenos Aires: Emecé, 1945.

______. *Historia general del Perú.* 3 vols. [1617]. Ed. A. Rosenblat. Buenos Aires: Emecé, 1944.

Geertz, Clifford. *The Interpretation of Cultures.* New York: Basic Books, 1973.

Gill, Sam. *Mother Earth: An American Story.* Chicago: University of Chicago Press, 1987.

______. *Native American Religious Action: A Performance Theory.* Columbia: University of South Carolina Press, 1987.

González Holguín, Diego de. *Vocabulario de la lengua general de todo el Perú* [1608]. Lima: Imprenta Santa María, 1952.

Goody, Jack. *The Domestication of the Savage Mind.* Cambridge: Cambridge University Press, 1977.

Gossen, Gary. *Chamulas in the World of the Sun: Time and Space in a Maya Oral Tradition.* Cambridge, Mass.: Harvard University Press, 1974.

Gow, David. "The Roles of Christ and Inkarri in Andean Religion." *Journal of Latin American Lore* 6:2 (1980):279–298.

Gow, Rosalind. "Inkarri and Revolutionary Leadership in the Southern Andes." *Journal of Latin American Lore* 8:2 (1982):197–223.

Guaman Poma de Ayala, Felipe. *El primer nueva corónica y buen gobierno.* [ca. 1613] 3 vols. Ed. J. Murra and R. Adorno. México: Siglo Veintiuno, 1980.

Hanke, Lewis. *The Spanish Struggle for Justice in the Conquest of America.* Philadelphia: University of Pennsylvania Press, 1949.

Harris, Olivia. "Complementarity and Conflict: An Andean View of Women and Men." In *Sex and Age as Principles of Social Differentiation*, pp. 21–40. Ed. J. S. LaFontaine. London: Academic Press, 1978.

Harrison, Regina. *Signs, Songs, and Memory in the Andes: Translating Quechua Language and Culture.* Austin: University of Texas Press, 1989.

Hernández Príncipe, Rodrigo. "Visitas de Rodrigo Hernández Príncipe" [1621–1622]. In *Cultura andina y represión: procesos y visitas de idolatrías y hechicerías, Cajatambo, siglo XVII*, pp. 461–507. Ed. P. Duviols. Cuzco: Centro de Estudios Rurales Andinos Bartolomé de Las Casas, 1986.

Hertz, Robert. "The Pre-eminence of the Right Hand: A Study in Religious Polarity." In *Right and Left: Essays on Dual Symbolic Classification*, pp. 3–31. Ed. R. Needham. Chicago: University of Chicago Press, 1973.

Howes, David. "Controlling Textuality: A Call for a Return to the Senses." *Anthropologica* 32:1 (1990):55–73.

_______. "The Shifting Sensorium: A Critique of the Textual Revolution in Anthropological Theory." Paper presented at the Twelfth International Congress of Anthropological and Ethnological Sciences, Zagreb, Yugoslavia, July 26, 1988.

_______, ed. *The Varieties of Sensory Experience: A Source Book in the Anthropology of the Senses.* Toronto: University of Toronto Press, 1991.

Isbell, Billie Jean. "The Metaphoric Process: 'From Culture to Nature and Back Again'." In *Animal Myths and Metaphors in South America*, pp. 285–313. Ed. G. Urton. Salt Lake City: University of Utah Press, 1985.

_______. "'Those Who Love Me': An Analysis of Andean Kinship and Reciprocity Within a Ritual Context." In *Andean Kinship and Marriage*, pp. 81–105, ed. R. Bolton and F. Mayer. Washington, D.C.: American Anthropological Association, 1977.

______. *To Defend Ourselves: Ecology and Ritual in an Andean Village.* Austin: Institute of Latin American Studies and University of Texas Press, 1978.

Isbell, William H. "Cosmological Order Expressed in Prehistoric Centers." *Actes du XLlle Congrès International des Américanistes.* Vol. 4, pp. 269–297. Paris; 1976.

Jackson, Michael. "Thinking Through the Body: An Essay on Understanding Metaphor." *Social Analysis* 14 (December 1983):127–149.

Johnson, Mark. *The Body in the Mind: The Bodily Basis for Meaning, Imagination, and Reason.* Chicago: University of Chicago Press, 1987.

Joralemon, Donald. "The Role of Hallucinogenic Drugs and Sensory Stimuli in Peruvian Ritual Healing." *Culture, Medicine and Psychiatry* 8 (1984):399–430.

Kubler, George. "The Quechua in the Colonial World." In *Handbook of South American Indians*, vol. 2: *The Andean Civilizations*, pp. 331–410. Ed. J. H. Steward. Bureau of American Ethnology Bulletin 143. New York: Cooper Square, 1963.

Kunzle, David. "World Upside Down: The Iconography of a European Broadsheet Type." In *The Reversible World: Symbolic Inversion in Art and Society*, pp. 39–94. Ed. Barbara Babcock. Ithaca, N.Y.: Cornell University Press, 1978.

Leach, Edmund. *Claude Lévi-Strauss.* New York: Penguin, 1970.

Lévi-Strauss, Claude. *From Honey to Ashes: Introduction to a Science of Mythology,* vol. 2. Trans. J. and D. Weightman. New York: Harper and Row, 1973.

______. *The Raw and the Cooked: Introduction to a Science of Mythology,* vol. 1. Trans. J. and D. Weightman. New York: Harper and Row, 1969.

______. *Tristes Tropiques.* Trans. J. Russell. New York: Atheneum, 1967.

Lira, Jorge A. *Medicina andina: farmacopea y ritual.* Cusco: Centro de Estudios Rurales Andinos Bartolomé de Las Casas, 1985.

López Austin, Alfredo. *The Human Body and Ideology: Concepts of the Ancient Nahuas.* 2 vols. Trans. T. and M. Ortiz de Montellano. Salt Lake City: University of Utah Press, 1988.

Lowe, Donald. *History of Bourgeois Perception.* Chicago: University of Chicago Press, 1982.

McDowell, John Holmes. *Sayings of the Ancestors: The Spiritual Life of the Sibundoy Indians.* Lexington: University Press of Kentucky, 1989.

McLuhan, Marshall. *Understanding Media: The Extension of Man.* New York: McGraw-Hill, 1964.

Mariscotti de Gorlitz, Ana María. "Los curi y el rayo." *Actes du XLlle Congrès International des Américanistes,* vol. 4, pp. 356–376. Paris, 1976.

Mauss, Marcel. "Body Techniques." *In Sociology and Psychology Essays,* pp. 95–123. Trans. B. Brewster. London: Routledge and Kegan Paul, 1979.

Means, Philip A. *Fall of the Inca Empire and the Spanish Rule in Peru: 1530–1780.* New York: Gordian Press, 1964.

Miles, Margaret. *Image as Insight: Visual Understanding in Western Christianity and Secular Culture.* Boston: Beacon Press, 1985.

Miracle, Andrew, and Juan Yapita Moya. "Time and Space in Aymara." In *The Aymara Language in Its Social and Cultural Context,* pp. 33–56. Gainesville: University of Florida Press, 1981.

Mishkin, Bernard. "The Contemporary Quechua." In *Handbook of South American Indians,* vol. 2*: The Andean Civilizations,* pp. 411–470. Ed. J. H. Steward. Bureau of American Ethnology Bulletin 143. New York: Cooper Square, 1946.

Molina "el Almagrista," Cristóbal de. "Relación de muchas cosas acaescidas en el Perú" [1552]. In *Crónicas peruanas de interés indígena,* pp. 57–95. Ed. F. Esteve Barba. Biblioteca de Autores Españoles, vol 209. Madrid: Ediciones Atlas, 1968.

Molina "of Cuzco," Cristóbal de. "Fábulas y ritos de los Incas" [1574]. In *Las crónicas de los Molinas,* pp. 7–84. Ed. F. Loayza. Lima: Los pequeños grandes libros de la historia americana, 1943.

Montesinos, Fernando de. "Memoriales antiguas historiales y políticas del Perú" [1644]. In *Colección de libros y documentos referentes a la historia del Perú,* ser. 2A, vols. 4–7. Ed. H. Urteaga. Lima: Sanmartí, 1911.

Morris, Brian. *Anthropological Studies of Religion: An Introductory Text.* Cambridge: Cambridge University Press, 1987.

Murra, John. "El 'control vertical' de un máximo de pisos ecológicos en las sociedades andinas." In *Visita de la Provincia de León de Huanuco,* vol. 1, pp. 383–406. Ed. Inigo Ortiz de Zúñiga. Huanuco, Peru: Universidad Hermilio Valdizan, 1972.

Murúa, Martín de. *Historia del origen y genealogía real de los reyes Incas del Perú* [1590]. Ed. C. Bayle. Madrid: Instituto Santo Toribio de Mogrovejo, 1946.

______. *Historia general del Perú: origen y descendencia de los Incas . . .* [1613]. Ed.

M. Ballesteros–Gaibrois. Madrid: Instituto Gonzalo Fernández de Oviedo, 1962.

Nash, June. *We Eat the Mines and the Mines Eat Us: Dependency and Exploitation in Bolivian Tin Mines.* New York: Columbia University Press, 1979.

Nuckolls, Janice. "Sound Symbolic Discourse as Cinematic Text." Paper presented at the annual conference of the American Anthropological Association, Chicago, 1991.

Oblitas Poblete, Enrique. *Cultura Callawaya.* La Paz: Talleres Gráficos Bolivianos, 1963.

Ocampo, Baltasar de. "Descripción y sucesos históricos de la provincia de Vilcabamba." In *Colección de libros y documentos referentes a la historia del Perú,* ser. 2A, vols. 4–7. Lima: Sanmartí, 1911.

Ong, Walter. *Orality and Literacy: The Technologizing of the Word.* London: Methuen, 1982.

______. *The Presence of the Word.* New Haven, Conn: Yale University Press, 1967.

______. "World as View and World as Event." *American Anthropologist* 71 (1969):634–647.

Ortiz Rescaniere, Alejandro. "El mito de la escuela." In *Ideología mesiánica del mundo andino,* pp. 238–243. Ed. J. Ossio A. Lima: Colección Biblioteca de Antropología, 1973.

Ossio A., Juan. *Ideología mesiánica del mundo andino.* Lima: Colección Biblioteca de Antropología, 1973.

Pachacuti Yamqui, Joan de Santacruz. "Relación de antigüedades deste reyno del Pirú." [ca. 1613]. In *Tres relaciones de antigüedades peruanas,* pp. 205–281. Ed. M. Jiménez de la Espada. Asuncion, Paraguay: Editorial Guaraní, 1950.

Pease, Franklin. *El dios creador andino.* Lima: Mosca Azul, 1973.

Pizarro, Pedro. *Relation of the Discovery and Conquest of the Kingdoms of Peru* [1571]. 2 vols. Ed. and trans. P. A. Means. New York: Cortés Society, 1921.

Platt, Tristan. "Mirrors and Maize: The Concept of *Yanantin* Among the Macha of Bolivia." In *Anthropological History of Andean Polities,* pp. 228–259. Ed. J. Murra, N. Wachtel, and J. Revel. Cambridge: Cambridge University Press, 1986.

Polo de Ondegardo, Juan. "Los errores y supersticiones de los Incas, sacadas del tratado y averiguación que hizo el Licenciado Polo." In *Colección de libros y documentos referentes a la historia del Perú,* ser. 1, vol. 3, pp. 3–43. Ed. H. H. Urteaga and C. A. Romero. Lima: Sanmartí, 1916.

______. "Instrución contra las ceremonias y ritos que usan los indios conforme al tiempo de su infidelidad" [1567]. In *Colección de libros y documentos referentes a la historia del Perú,* ser. 1, vol. 3, pp. 189–203. Ed. H. H. Urteaga and C. A. Romero. Lima: Sanmartí, 1916.

______. "Relación de los fundamentos acerca del notable daño que resulta de no guardar a los indios sus fueros" [1571]. In *Colección de libros y documentos referentes a la historia del Perú,* ser. 1, vol. 3, pp. 44–188. Ed. H. H. Urteaga and C. A. Romero. Lima: Sanmartí, 1916.

Porras Barrenechea, Raúl. "Prólogo." In *Antología del Cuzco,* pp. iii–xxxviii. Ed. R. Porras Barrenechea. Lima: Librería Internacional del Perú, 1961.

Riva–Agüero, José de la. *Obras completas de José de la Riva–Agüero,* vol. 5. Lima: Pontificia Universidad Católica del Perú, 1966.

Rostworowski de Díez Canseco, María. *Pesos y medidas en el Perú pre-hispánico.* Lima: Imprenta Minerva, 1960.

Rowe, John H. "The Age-Grades of the Inca Census." In *Miscellanea Paul Rivet,* vol. 2, pp. 499–522. Mexico: Trigésimo primero Congreso Internacional de Americanistas, Universidad Nacional Autónoma de México, 1958.

______. "Inca Culture at the Time of the Spanish Conquest." In *Handbook of South American Indians,* vol. 2: *The Andean Civilizations,* pp. 183–330. Ed. J. H. Steward. Bureau of American Ethnology Bulletin 143. New York: Cooper Square, 1963.

______. "What Kind of Settlement was Inca Cuzco?" *Ñawpa Pacha* 5 (1967):59–76.

Santillan, Hernando de. "Relación del origen, descendencia, política y gobierno de los Incas" [1563]. In *Crónicas peruanas de interés indígena,* pp. 97–149. Ed. F. Esteve Barba. Biblioteca de Autores Españoles, vol. 209. Madrid: Ediciones Atlas, 1968.

Santo Tomás, Domingo. *Léxicon o vocabulario de la lengua general del Perú* [1560]. Lima: Universidad Nacional Mayor de San Marcos, 1951.

Sarmiento de Gamboa, Pedro. "Historia Índica" [1572]. In *Obras de Garcilaso de la Vega,* vol. 4, pp. 195–279. Ed. C. Sáenz de Santa María. Biblioteca de Autores Españoles, vol. 135. Madrid: Ediciones Atlas, 1965.

Seeger, Anthony. "The Meaning of Body Ornaments : A Suya Example," *Ethnology* 14 (1975): 211–224.

______. *Nature and Society in Central Brazil: The Suya Indians of Mato Grosso.* Cambridge, Mass.: Harvard University Press, 1981.

Sharon, Douglas. *Wizard of the Four Winds: A Shaman's Story.* New York: Free Press, 1978.

Silverblatt, Irene. "The Evolution of Witchcraft and the Meaning of Healing in Colonial Andean Society." *Culture, Medicine and Psychiatry* 7 (1983): 413–427.

______. *Moon, Sun, and Witches: Gender Ideologies and Class in Inca and Colonial Peru.* Princeton, N.J.: Princeton University Press, 1987.

Silverman-Proust, Gail. "Weaving Technique and the Registration of Knowledge in the Cuzco Area of Peru." *Journal of Latin American Lore* 14:2 (1988):207–241.

Stark, Louisa. "The Lexical Structure of Quechua Body Parts." *Anthropological Linguistics* 2 (January 1969):1–15.

Stoller, Paul. *The Taste of Ethnographic Things: The Senses in Anthropology.* Philadelphia: University of Pennsylvania Press, 1989.

Sullivan, Lawrence. "Sound and Senses: Toward a Hermeneutics of Performance." *History of Religions* 26:1 (1986):1–33.

Thrupp, Sylvia, ed. *Millennial Dreams in Action: Studies in Revolutionary Religious Movements.* New York: Schocken Books, 1970.

Titu Cusi Yupanqui, Diego de Castro. *Ynstrución del Ynga don Diego de Castro Titu Cussi Yupanqui* [1570]. Ed. Luis Millones. Lima: Ediciones El Virrey, 1985.

Todorov, Tzetan. *The Conquest of America: The Question of the Other.* Trans. R. Howard. New York: Harper and Row, 1984.

Tschopik, Harry, Jr. "The Aymara of Chucuito, Peru." *Anthropological Papers of the American Museum of Natural History* 44 (1951):133–308.

Turner, Bryan. *The Body and Society.* Oxford: Basil Blackwell, 1984.

Urbano, Henrique-Osvaldo. "Dios Yaya, Dios Churi y Dios Espíritu: modelos trinitaríos y arqueología mental en los Andes." *Journal of Latin American Lore* 6:1 (1980):111–127.

______. *Wiracocha y Ayar: héroes y funciones en las sociedades andinas.* Cuzco: Centro de Estudios Rurales Andinos Bartolomé de Las Casas, 1981.

Urioste, George. "Sickness and Death in PreConquest Andean Cosmology." In *Health in the Andes*, pp. 9–318. Ed. J. Bastien and J. Donahue. Washington, D.C.: American Anthropological Association, 1981.

———, ed. *Hijos de Pariya Qaqa: la tradición oral de Waru Chiri* [ca. 1608]. 2 vols. Syracuse, N.Y.: Syracuse University Maxwell School of Citizenship and Public Affairs, 1983.

Urton, Gary. "Animal Metaphors and the Life Cycle in an Andean Community." In *Animal Myths and Metaphors in South America*, pp. 251–284. Ed. G. Urton. Salt Lake City: University of Utah Press, 1985.

______. *At the Crossroads of the Earth and the Sky: An Andean Cosmology.* Austin: University of Texas Press, 1981.

______. *The History of a Myth: Pacariqtambo and the Origin of the Incas.* Austin: University of Texas Press, 1990.

______ and R. T. Zuidema. "La constelación de la Llama en los Andes peruanos." *Allpanchis* 9 (1976):59–119.

Vaca, Cristóbal de. "Discurso sobre la descendencia y gobierno de los Incas" [1542]. In *Colección de libros y documentos referentes a la historia del Perú*, ser. 2, vol. 3. Ed. H. Urteaga. Lima: Sanmartí, 1920.

Valencia Espinoza, Abraham. "Inka Qollari dramatizado." In *Ideología mesiánica del mundo andino*, pp. 283–298. Ed. J. Ossio A. Lima: Colección Biblioteca de Antropología, 1973.

Vinge, Louise. *The Five Senses: Studies in a Literary Tradition.* Lund, Sweden: Royal Society of Letters at Lund, 1975.

Wachtel, Nathan. *The Vision of the Vanquished: The Spanish Conquest of Peru Through Indian Eyes, 1530–1570.* Trans. B. and S. Reynolds. New York: Barnes and Noble, 1977.

Wagner, Catherine Allen. "Coca, Chicha and Trago: Private and Communal Rituals in a Quechua Community." Ph.D. diss. University of Illinois, Urbana-Champaign, 1978.

Zuidema, R. T. *The Ceque System of Cuzco.* Leiden: E. J. Brill, 1964.

______. "The Inca Calendar." In *Native American Astronomy*, pp. 221–226. Ed. A. F. Aveni. Austin: University of Texas Press, 1977.

______. "The Inka Kinship System: A New Theoretical View." In *Andean Kinship and Marriage*, pp. 240–292. Ed. R. Bolton and F. Mayer. Washington D.C.: American Anthropological Association, 1977.

______. "Inka Dynasty and Irrigation: Another Look at Andean Concepts of History." In *Anthropological History of Andean Polities*, pp. 177–200. Ed. J. Murra, N. Wachtel, and J. Revel. Cambridge: Cambridge University Press, 1986.

______. "The Lion in the City: Royal Symbols of Transition in Cuzco." *Journal of Latin American Lore* 9:1 (1983):39–100.

______. "Masks in the Incaic Solstice and Equinoctial Rituals." In *The Power of Symbols: Masks and Masquerade in the Americas*, pp. 149–156. Ed. N. R. Crumrine and M. Halpin. Vancouver: University of British Columbia Press, 1983.

______. "Myth and History in Ancient Peru." In *The Logic of Culture: Advances in Structural Theory and Methods*, pp. 150–175. Ed. I. Rossi and contributors. London: Tavistock, 1982.

______. "Shafttombs and the Inca Empire." *Journal of the Steward Anthropological Society* 9 (1978):133–178.

INDEX